Transports of Delight

Peter Hancock

Transports of Delight

How Technology Materializes Human Imagination

 Springer

Peter Hancock
Department of Psychology
University of Central Florida
Orlando, FL
USA

ISBN 978-3-319-85607-0   ISBN 978-3-319-55248-4   (eBook)
DOI 10.1007/978-3-319-55248-4

Printed on acid-free paper

This Springer imprint is published by Springer Nature
The registered company is Springer International Publishing AG
The registered company address is: Gewerbestrasse 11, 6330 Cham, Switzerland

Praise for the Text

"Hancock's *Transports of Delight* is a journey back through time to discover the future. The book artfully illustrates the theme that in creating our technologies we are re-creating ourselves, leaving us with the question of whether we are climbing higher, traveling an unending circular labyrinth, or regressively falling into predictable catastrophes?"

Professor John Flach – Wright State University. Author of "*What Matters.*"

In this book Peter Hancock wonders at wonderment, he has insights into insightfulness, and is curious about curiosity. Peter is our historical journeyman who guides us on his voyage of discovery about the essence that drives humankind. The parody of the journey is not lost on us. The overwhelming desire to create and consume seems to lead to the inevitable consequence of destruction. Can we not learn the lessons of history mapped out before us? It seems that they are all too quickly lost in the mist of time, only to be relearnt by successive generations following similar paths. Humankind seems destined to create structures and effects that are far greater than ourselves and, as a direct consequence, become the architects of our own destruction. There is something of exasperation in Peter's melancholy tone. If we are to break the familiar cycle of destruction littered across history, then we should heed his words. To be blunt, reading this book could literally save humankind. But that would assume we could actually follow the path of Peter's curiosity, wonder and insights. I fear we may not.

Professor Neville Stanton, University of Southampton. Author of "*Digitizing Command and Control.*"

"Almost every area of human endeavor leads to aspiration-fueled envelope-pushing. Human beings seem destined to create catastrophic self-deceptions by believing that bigger is better. Peter Hancock has led us, philosophically, through a garden of aspiration-fueled constructions whose envelopes tore from overstuffing. This book is at once an enjoyable read, a thought-provoking thesis, and a history lesson."

Professor John Senders, Professor Emeritus – University of Toronto, "*Sapientissimo hominum*"

"*Peter Hancock's Transports of Delight adroitly conveys the spiritual dimension of mankind's great technological constructions, including both the glory and the sometime tragedy. It transports us in our imaginations to a higher place. This is a superb contribution, beautifully written and a treat to read.*"

Professor Tom Sheridan, Professor Emeritus – Massachusetts Institute of Technology (MIT) and Author of "*What Is God? Can Religion Be Modeled?*"

"*Transports of Delight - Here is human achievement, in its glories and its sorrows. It is common to invoke our sense of wonder over the contents of the natural world, but in this book Peter Hancock shows that the man-made world of technological achievements is as full of wonders as is the natural world. A most unusual and thought-provoking book.*"

Neville Moray, Professor Emeritus, University of Surrey and Author of "*Science, Cells, and Souls.*"

Ruminations on the nature of the constructions through which human beings transport themselves to other realms; their fabrication, their meaning and their destruction, with an examination of their similarities and differences, and the way in which their legends are interwoven into the narrative consciousness of all humankind.

"Even for the atheist and rationalist, there are places in the world that are special, for no reason that can be easily explained"

Pears, I. (2002). *The dream of Scipio* (p.263). Riverhead, N.Y.

Table of Contents

ACKNOWLEDGMENTS

Many people have helped to make the present text a reality. First and foremost I must acknowledge the debt I owe to all of my students. Many have personally helped in the present research and production process, and even those who have not had a direct hand in the present production have had effects on me, for which I am continuingly grateful. With respect to specific students, Theresa Kessler's devotion to the present project was unparalleled. I am also grateful to her husband Richard Kessler for creating several of the present illustrations. I am especially indebted to Angela Bardwell-Owens for her assistance in the various phases of development of the work and to Keith MacArthur whose consistent efforts were very much appreciated throughout. In addition, thanks go to Mitchell Dunfee who assisted, especially in relation to the present illustrations. My daughter Gabriella Hancock proof-read and edited the whole manuscript as she has done with a number of my books; her help is much appreciated. My wife Frances is a constant source of love and encouragement to me; my debt to her is profound and, as yet, unredeemed. She also acted as a stern critic of the work and of me, which spouses are destined to be.

In terms of thanks to those who have provided me with intellectual and academic mentoring over the years, I remain deeply grateful to Karl M. Newell my advisor, mentor and friend as well as to Joel S. Warm, colleague and guide. I stand indebted to, and in the shadow of, Thomas B. Sheridan, Neville P. Moray, Raja Parasuraman and John Flach. Their unstinted guidance and wisdom across the years has always been a source of motivation and inspiration. Finally I stand in the debt of, and in awe of, John W. Senders who has always been simply, and uncompromisingly

– John. I am further especially grateful to the numerous librarians and museum curators who have provided access to important original materials for this text. Most of all, I am thankful to those who labor to keep Gothic Cathedrals open and flourishing. Theirs is a hard but highly praiseworthy endeavor. They are sentinels of time and custodians of some of our greatest human treasures. I dedicate this work to them.

"the ultimate purpose of technology is to 'transport'
its patrons to other realms of existence."

PREFACE

HISTORY IS SIMPLY OUR FUTURE IN REVERSE

History is our collective memory. History mirrors our own personal memory in that it is selective, fallible, and perishable [1]. Like our individual memory, history acts to record, highlight and recall events that are especially intense, unusual, or meaningful to those who witness them [2]. In this way, history proves to be a flawed and subjective record that omits much of what is seen and heard in favor of recalling the striking patterns found in the few 'unforgettable' events. Efforts to capture history as an extended stream of consciousness are well-meaning and well-intentioned. They recognize that the vast majority of the past is made up of mundane everyday happenings that compose the basic fabric of existence. However, like the mundane in each of our own personal lives, the lives of the mundane are, sadly, largely consigned to the corridors of forgotten time. All the insistence in the world on the vitality of each and every individual does not render our own personal memory, or indeed our history, any different. We are what we remember. As the totality of collective memory, history is composed then of a few brief shining moments. This book tells a story which weaves the threads that tie some of those very special moments together.

Some have argued that there is no such thing as history, only biography. History, from this perspective, represents a collection of the stories of people scattered across time. By this measure, geography would represent those self-same stories distributed across space [3]. Interestingly, the basic concepts of space and time are themselves truly indivisible; that is, as opposed to the way we commonly perceive them and formally parse them on a day to day basis. In commenting on the inseparability of space and time at the turn of the twentieth century, the German physicist Herman Minkowski asked, *"who has been at a place except at a time, and who has experienced time except at a place?'* He predicted that: *"Henceforth, space by itself and time by itself are doomed to fade away into mere shadows and only a union of the two will preserve an independent reality"* [4]. The nominal division then between human history and human geography should, from Minkowski's perspective, gradually dissolve. In the future we will perhaps not even use separate names for such studies. What we see are visions across a time-scape. I want to present and articulate such vision here.

One of the most basic functions of the human mind is to distill patterns from, and/or impose patterns upon, these respective visions [5]. Sometimes the framing pattern is supposedly 'obvious' and results directly from the way we are genetically wired to perceive objects in space and acknowledge their continued existence in time. The marvelous insight of the theory of evolution for example, was to see that entities which previously we had been calling by different names could actually change or transform into each other on a time scale which far exceeded the lifetime of any single human observer [6]. Thus, it was Darwin who was truly *'the man who saw through time.'* [7] This insight, which represents Darwin's personal act of imagination,

Figure P-1: *The descendant of the tree under which Newton is purported to have been struck by the 'apple' of gravity. Woolthorpe Manor (Photograph by the Author).*

was itself spread across a span of years unlike that of Wallace to whom the same vision was vouchsafed in an episode of acute fever. Yet each of us could have envisaged same process of change. For example, we see our own children change from newborn to infant, and then mature from adolescent into adult. There was no apparent mystery in this directly perceived process. But in western society, we do not call an adult by a different name from when they were a child. [8] Darwin's insight was to 'envision' this same evolving pattern in a landscape of change across nominally different organisms and resultant species, on a time scale that no one human could ever directly 'see' happening. This was his 'unique' vision and was one, I have argued, that was based upon a different cultural perspective of time [9]. Isaac Newton had also previously experienced the self-same brisance of insight. The commonplace anecdote is that Newton saw a common influence on a falling apple (from his own apple tree and see Figure P-1), and the comparable effect on an ever-falling (i.e., orbiting) moon [10]. By understanding this commonalty, Newton was subsequently able to provide a mathematical description of ubiquitous gravitational effects. While perhaps doubting the veracity of this particular story, we rightly applaud Newton for his staggering and illuminating revelation.

The same type of insight had previously been rendered to Johannes Kepler who, convinced of God's perfection, sought the mathematical simplicity which he "knew" must underpin the orbits of the so-called 'wandering stars,' (the Planets) [11]. When he at last found and articulated the elliptical key to the paths of the orbits of the planets, it was for him very much a 'Eureka' moment of spiritual fulfillment, as all such moments necessarily are [12]. A profound vision induces transcendence by definition – they transcend. Any such moment of vision is always a very special occasion [13]. In a similar manner, we can identify circumstances in which we recognize the special insights of poets and artists, as well as philosophers and scientists, who take an unusual concatenation of conditions (or views across the landscape of space-time) and articulate what they see or feel into larger principles [14]. Such moments of cognitive ecstasy literally elucidate our collective understanding of our world and our place within it.

The overarching question with respect to the "landscape" itself then is this: how far are all things different and how far are all things the same? Are things actually random and we humans only impose our patterns upon such randomness? Or, in contrast, are all things connected and our limited powers of perception only slowly and haphazardly reveal the full articulations of these reticulated connections? Or could it be that this is actually a false division, splitting apart a person from their environment which happens only because humans are so uniquely self-centered? To some extent, of course, we are also constrained to see this as a question of how far any fracturing of common perception or received wisdom is the act of genius or the product of insanity. For, of course, both genius and insanity are necessarily similar in being so profoundly far from that which is 'normal' [15].

Coherent visions (one might even say theories of being) necessarily represent links between known things. One cannot experience such a vision on the basis of what one doesn't know (i.e., that which is understood either explicitly or implicitly). However, when such a vision strikes, its power is assessed by the way it illuminates whole vistas at the boundary between our knowledge and our ignorance. Newton's insight was so profound that much of the then known universe was exposed by the lighting of his conception [16]. Rarely do illuminations of such an elemental nature and order of magnitude occur. Yet all of us, in individual ways accomplish this selfsame process of illumination everyday of our lives.

For children the process of, and instances of, wonder happen frequently because of the range of things they are coming to know but have not yet connected together. Even the most jaded pedagogue can communicate staggering visions to children simply through reference to already established insights that the child has yet to grasp. [17]. Unfortunately, for most adults these moments of wonder occur progressively less frequently. This effect results largely from the range of things that they think they already understand! It seems that most people get to a point in their life where they believe they know enough to conduct their lives satisfactorily and then tragically, venture no further [18]. Few people are fortunate enough professionally to pursue such wonder on a daily basis. It's not that our capacity for wonder is ever totally extinguished; it is simply that like any other capability, it becomes dormant and atrophies with lack of use.

To be truly profound, wonder has to be self-generated [19]. But today, our society relies so much on the media and the entertainment industry to provide pallid shadows of this self-generated wonder that we often become passively accepting of others' impositions on us. The present text is my attempt to engage in active wonder. In so doing I look to encourage readers to participate in this process in their own special and unique way. If the patterns I explore here interests readers I will be gratified. However, if it stimulates them to explore their personal visions I will be fulfilled indeed. For each and every one of these moments of human wonder are *'Transports of Delight.'*

Reference Notes: Preface

[1]. Schacter, D. (2001). *The seven sins of memory.* Houghton-Mifflin: New York.

[2] See, e.g.: www.eyewitnesstohistory.com/

[3]. I know that this statement will enrage many physical geographers, geologists, and astronomers who have devoted their whole personal and professional lives to the study of the distribution of objects in space-time. However, they should recognize that the human animal is certainly the most self-involved organism on Earth and perhaps holds even a universal record in this respect. I hope the qualifying appellation 'human' will temper their anger, if only for a short interval of space-time.

[4]. Minkowski, H. (1908/1923). Space and time. In: H.A. Lorentz, A. Einstein, H. Minkowski. and H. Weyl (Eds). *The principles of relativity.* London: Dover.

[5] When the vision is truly an enlightening one we call it an act of genius. However, people see patterns all the time, we often see false faces in non-face patterns; a tendency known as pareidolia. Our general tendency to impose nominally 'false patterns on perceptual distributions is called apophenia and certain patterns may be the result of cognitive apophenia. But a central question remains whether the arbiter of the 'correctness' of these patterns, i.e., society, is actually right in saying what patterns are 'false' and what are 'real.'

[6]. It is most interesting to note that rarely do human cultures actually adopt this atemporal way of thinking in which the name of an object or entity changes as the object or entity itself varies over time. Thus some natives of the Marquesas in the Pacific (Fatu-Hiva) have different words for the same fruit in different stages of ripeness from green fruit through ripeness to a rotten state. It's intriguing to speculate, that Islanders whom Darwin met in Tahiti, may have communicated this to him. It is seductive to think that such an atemporal perspective could have triggered Darwin's notion of progressive change by understanding that all naming is arbitrary and thus radical change is possible over time even if the name remains the same. See: Hancock, P.A. (2007). On time and the origin of the theory of evolution. *Kronoscope*, 6 (2), 192-203.

[7] I use the phrase advisedly here even though Eiseley's delightful text actually refers to Francis Bacon, for whom a very good case can also be made. See; Eiseley, L. (1973). *The man who saw through time*. Scribner: New York. The comparative stances of Darwin and Wallace and their shared, but independently discovered vison, are one of the great stances of science, van Wyhe, J. (2013). *Dispelling the darkness: Voyage in the Malay archipelago and the discovery of evolution by Wallace and Darwin*. World Scientific Publishing Co.: Singapore.

[8] Of course there are cultures in which the rate of passage into adulthood involves exactly such a re-naming process and in our own society some groups even talk of being "reborn" on occasion. Also, western children do have 'nicknames' which are frequently dropped in adulthood.

[9] Of course, Darwin's vision was not unique as Wallace's re-capitulation, or essentially co-discovery, clearly showed.

[10] This may well be one of those wonderful legends that does not actually represent reality.
See: http://csep10.phys.utk.edu/astr161/lect/history/newtongrav.html

[11] Koestler, A. (1959). *The sleepwalkers: A history of man's changing vision of the universe.* Penguin: London: England. Kepler's work here is predicated upon Brahe's extensive empirical observations.

[12] The precise term should be 'transcendental' since spiritual still carries an overtone of formal religion. And see also: Topolinski, S., & Reber, R. (2010). Gaining insight into the "Aha"-experience. *Current Directions in Psychological Science, 19,* 402-405.

[13] For an account of how Friedrich August Kekulé discovered the cyclic structure of Benzene see: http://web.chemdoodle.com/kekules-dream. It is rather interesting to note that many of the great insights that have occurred to people have done so when they are unnaturally hot. Consider Descartes and his sojourn in the sauna, Wallace and his discovery of evolution in a fever, and even Biblical precedents such as the burning-fiery bush!

[14] Two of my favorites in this respect are Dali and Magritte: see, http://thedali.org/home.php; and also: http://www.musee-magritte-museum.be. Neither artist was an outstanding painter per se, yet both generated and represented some truly startling human visions.

[15] see: Hancock, P.A. (2012). Notre trahison des clercs: Implicit aspiration, explicit exploitation. In: R.W. Proctor, and E.J. Capaldi, (Eds.), *Psychology of science: Implicit and explicit reasoning.* (pp. 479-495), New York: Oxford University Press.

[16]*"Nature and nature's laws lay hid in night: God said, 'let Newton be!' and all was light."* (Alexander Pope, 1688-1744). See also: Fauvel, J., Flood, R., Shortland, M., & Wilson, R. (1988), (Eds.). *Let Newton be! A new*

perspective on his life and works. Oxford University Press: Oxford. See also: Glecik, J. (2003). *Isaac Newton.* Vintage: New York.

[17] One can achieve this easily by proving the Pythagorean theorem through the use of geometric shapes contained within a square (see Bronowski, J. (1973). *Ascent of Man.* BBC Books: London). And see also the series by James Burke which represents television as the best it can be: Burke, J. (1978). *Connections.* Little, Brown & Co.: New York.

[18] One might say that people generally 'satisfice' their life, as opposed to 'optimizing' it, in terms developed by the Nobel-prize winning scientist, Herbert Simon. And see: Simon, H.A. (1996). *The sciences of the artificial.* (3rd ed.). MIT Press: Cambridge, Mass.

[19] However wonderful any communicated insight can be, there is a necessary distinction between self and non-self, as well as a crucial difference between original self-discoveries, as opposed to the general category of non-original but communicated observation. And see: Polanyi, M. (2012). *Personal knowledge: Towards a post-critical philosophy.* University of Chicago Press: Chicago.

"Ravishing. Yes. It is the word I always use to describe it. It transports one, does it not. Almost forcibly, carrying one rapturously away, to another and better world. What would it be to live there! One would never wish to leave." [1]

1. INTRODUCTION

ONE NARRATIVE OF MANY STORIES

1.1. Proem

My book is not just an historical account of each of the respective *'transports'* that I describe. In fact, their creation, their existence and their destruction have been documented elsewhere, and in more detail than I accomplish here [2]. For my present purposes, I have drawn liberally from these various sources and I am happy to acknowledge my profound debt to each of them. Since my intention is not to try explicitly to elaborate upon each factual corpus of the studies of any of these monuments, my initial three chapters represent a narrative summary of each of the stories of the respective *'transports'* I have selected. However, it is from these critical, individual foundations that I subsequently look to explore ever further into 'higher' synthetic, semiotic, synergistic, and symbiotic connections.

Before presenting each of these stories I want to begin by giving the reader an idea of how this whole present enterprise began. It is the story of a series of my own *'moments of wonder.'* My personal story started with my one-time membership of the *Titanic Historical Society*. I had long been interested in this most discussed, debated, and disputed of all human disasters well

© Springer International Publishing AG 2017
P. Hancock, *Transports of Delight*, DOI 10.1007/978-3-319-55248-4_1

before the more modern interest as characterized in James Cameron's motion picture [3].

The sinking of the *Titanic* was the overture to the cataclysm of the First World War, the so-called 'war to end all wars,' which of course it didn't. When *Titanic* sank, the myopic arrogance of the late-Victorian and early-Edwardian world-view was violently shaken. It was only later on the fields of northern France where all such certainty was definitively and finally swept away [4]. It was where the new arts of war had raised killing to an industrial level. Whatever our contemporary arguments about the impact of relativistic concepts derived from quantum physics, cultural relativism, and philosophical and artistic post-modernism, it was at *Paschendale*, *Ypres*, and *Vimy Ridge* that the comforting view that *'God's in his heaven, all's right with the world,'* [5] was explosively dismissed; at least from modern Western consciousness. Now, almost no one can recall from personal experience that horrific obscenity when nineteenth century military tactics met twentieth century killing technology. Few today even realize that more men were killed and injured on the first morning of the *Somme* offensive than were lost by the United States in the whole of the Viet Nam conflict [6]. Truly, it was the quintessential slaughterhouse of certainty.

1.2. Embarking on a Personal 'Voyage.'

Titanic then was the first shiver, the first frisson, the first suspicion that technology was uncontrollably outstripping the contemporary level of comprehension and control. The fact that the United

Kingdom's Board of Trade rules did not insist on lifeboat places for everyone on board was simply a reflection of the same misperception of technical expansion that produced the charges of horse cavalry into machine guns just two years after *Titanic* sank [7]. Society had, at that moment in time, developed a power that it could not fully control. Perhaps the cyclic tension between technical advance and insufficient social adjustment is a necessarily recurring one [8]?

So, the *Titanic* disaster, as well as being interesting with respect to the specific events of April 13th and 14th 1912, is also memorable as the quintessentially symbolic event of the early twentieth century. Indeed, I believe it is actually this symbolism, as much as the reality of the night itself, that rightly keeps it fresh in our thoughts. In the cold flat calm of the north Atlantic, the apparently impossible happened; the 'unsinkable' ship sank. So, my personal voyage began with *Titanic* and its now sadly clichéd *'Night to Remember'* [9]. However, *Titanic* was only part of the puzzle for which many more fertile seeds were needed to resolve the overall mystery. Recognition of the significance of the *Titanic* then was a necessary but not sufficient condition for fuller insight. *Titanic* is a graphic and magnificent story, but not one of my own personal 'experience.' The first story from my own personal perspective, which sent me moving down the present road, happened some miles south-west of Paris, for there lies the Cathedral town of Chartres [10].

1.3. The Wonder That Is *Chartres*

I first saw *Chartres* when I was in my early thirties - it is a truly stunning sight. Having traveled quite extensively in conjunction with my work, I had become progressively more dismayed by the evident convergence of diverse human environments under the driving force of our modern trans-global culture. It appeared to me as though fast-food franchises and multi-national commercial enterprises were growing like a global virus. Through media infiltration and commercial exploitation, a common culture is indeed beginning to invade widely diverse and separate human ecologies [11]. In just one poignant example, Harvard House, in Stratford-upon-Avon, in England, which was the house from which the founder of Harvard University had emigrated to America, was at one time a Pizza Hut!

As with many human endeavors, some of this convergent activity might be considered to be beneficial although in the present context of wonder, the majority appear to be destructive. Untrammeled, mean-spirited, greedy exploitation is often condemned as contemptible. Nevertheless, untrammeled, mean-spirited, greedy exploitation may well describe the process of evolution itself. However, it remains the case that the wonder and the vision that I am talking of, like evolution itself, need variation and require difference. Especially the presence of objects, events, and artifacts that are diverse and expansive not common and constrained. For 'wonder,' we cannot have new 'visions' if our experience is packaged, sanitized, curtailed, standardized, and unvarying. We require the unusual and the special, and *Chartres* is precisely that; it is truly unique.

I do not consider myself an especially spiritual or mystical being, but '*dull would he be of soul who could pass by a sight so touching in its majesty*' [12]. *Chartres* Cathedral can be seen from afar and has a startling impact as it appears above the surrounding fields. My first sight of the cathedral was from just such a panorama which was accompanied by few, if any, surrounding objects or structures through which to secure any impression of its size. So, my initial glimpse proved to be a visual illusion. The cathedral actually appeared to me to be a small village church, and one which was very close by. This illusion was only dispelled as I slowly began to realize that, even approaching at the speed of a fast automobile, there was no obvious change in the appearance of the building's size [13]. My first moment of wonder then, which connects all the themes of the present text together, was this instant when I first recognized the presence of the cathedral and felt the sensation of its massive size at a distance. A second impression, which quickly followed, was that *Chartres* Cathedral was like a giant ship sailing on a land-based ocean (Figure 1-1) [14]. It appeared literally to float above the surrounding countryside like some '*titanic*' voyager. Figure 1-1 barely gives even a shadowy reflection of this perceptual phenomenon. Nothing can compare with the experience of observing this at first hand. I strongly advocate that you look to do so for yourself. It was this observation that first prompted me to see the cathedral actually as a form of 'transport.'

Phenomenally, *Chartres* Cathedral does not disappoint upon closer inspection. Having conceived the idea that the impact of *Chartres* was one that was made at a distance, I was stunned to find the wonder multiplied upon itself upon closer approach. The first time I visited, I parked alongside the river and had to climb up the ridge upon on which the cathedral stands. It progressively imposes itself upon your sense of perception as you expend the physical effort to ascend the hill. Through the Royal Portal and into *Chartres'* vast, dark, wondrous cave, one knows instinctively that you are in the presence of something transcendental. From the

Figure 1-1. A picture of Chartres Cathedral across a 'sea' of corn. Sailing as majestically as an ocean-going Liner. (Illustration taken from the website: http://www.chartres-csm.org/). From this image, one can scarcely believe the Cathedral actually sits in the center of a bustling town.

deceptively slight but intriguing slope of the floor itself as it climbs beyond the main entrance into the heart of the nave, to the fabulously piquant labyrinth incised into the Cathedral's floor, from the brutally amazing stained-glass windows to the high altar itself; all architectural features combine to bombard and even overwhelm the senses of the unprepared visitor [15]. The bare simplicity of much of the chancel contrasts with the intricate screen work, the outside decoration with statuary and the asymmetry of the external towers. These are all characteristics and contrasts that once seen can never be forgotten. *Chartres* is a moment of wonder which is frozen in time. It was upon this, my first visit, that some inkling of the content of my present narrative were first revealed [16].

1.4. *Rheims* Not *Chartres*

Having eulogized *Chartres*, I must now turn to a curious quirk in respect of my evolving story. It was not at *Chartres*, but at *Rheims* where the next link in my progress was forged [17]. I had been on a visit to Spain with my family and our return route to England took us toward Calais from southern France. Our overnight stop was Rheims. We had arrived at our hotel very late at night and expected to resume our travel in the morning, expectedly in time to meet our appointed cross-channel ferry. When I woke, I tried to persuade my still tired family that we had just enough time to visit *Rheims* Cathedral. I was advised, in no uncertain terms, that I might want to make that visit a solitary perambulation! So at just past five a.m. on a spectacularly sunny summer's morning in northern France, I walked alone into the square in front of *Rheims'* Cathedral. A café-au-lait just before six a.m. saw me, a sentinel, awaiting the opening of the Cathedral portal. I was the first and only individual to enter after the turnkey had done his duty and I progressed in splendid isolation up the magnificent nave. I was just looking back at the simply incredible Rose Window when the organ burst forth in all its power and majesty [18]. The experience provided to me my next moment of wonder.

With no one around, I was momentarily disoriented in time. I understood that for those who were the first to worship in this cathedral, shortly after it was completed, it was not simply a center of religion; it was the center of everything. It was the source of the news, of entertainment, the site of most social gatherings. In fact the cathedral was the very epicenter of their world and their life. More, it was the *height* of technology. What I was looking at was a technical wonder that, because it had proved so durable and lasting, had at last rendered itself truly anachronistic. And now we, who are simple, passing transient beings, can see it only as a relic, an antique. *Rheims* Cathedral is a structure which is no longer the latest and greatest, as it was at the time of its inauguration [19]. Now putatively, and because of the myopia of human comprehension, it is an historical monument.

In respect of modern technology, we now think in terms of life cycles in which technology's conception, design, and fabrication with its subsequent redundancy, destruction and disappearance is expected to happen not merely within a lifetime but now perhaps just a decade. And now even more recently these technical cycles of electronic goods happen even within a year, a month, or for software a progressively vanishing moment. Such is the transience of modern celebrity – both human and technical [20]. Cathedrals are not seen as the height of technology because they have resisted this demise and lasted well beyond any individual human lifetime, they are the millennial technology. Like Darwin's fossils they litter our landscape but truly they remain wonders that are rendered material for all to see [21].

How fickle then, but also how paradoxically personal even social memory can be. I can illustrate this point rather easily. One merely has to ask an individual about their reaction to the assassination of President John Fitzgerald Kennedy. For those who were alive at the time it is an unequivocally visceral response to reflecting feelings engendered at the tragic moment itself [22]. For those not yet born then, it is simply an historical occurrence. A comparison might be the death of Princess Diana [23]. For individuals who can remember hearing of her death in Paris, I can tell them unequivocally that there is already a rising generation who now ask in all naiveté 'Who was Princess Diana?' If you don't feel the point I am trying to make through these two specific examples, you certainly will in your future. And now even the Princess Diana example is becoming dated; especially in light of the more recent death of other famous celebrities such as Michael Jackson [24]. This traveling window of individual and social consciousness will ever continue. The only thing that apparently changes is your position in time with respect to each occurrence. The basic rhythm of life itself remains untouched.

1.5. Too Successful a Technology?

Recognizing the amazing longevity of, and the poignancy of, these Gothic structures was my personal next step along my own journey. Like the less dramatic, but nevertheless poignant passing of the great ocean liners, I searched among these cathedrals for a 'Titanic' disaster. As we shall see I quickly found my 'shipwrecked' cathedral. Technologies come and technologies go and their wreckage is strewn everywhere about us in time as well as in space. I soon then had two of my transports, one of earth and one of water. However, it was time to look for the third and final member of the trinity that I required. My investigations of the *Hindenburg* were initially then an extension of my continuing vision. Part an interest in technological and human disasters and also part an interest in the communication of momentous events, the study of the *Hindenburg* was fresh in my mind as I pondered an inherent linkage between the four elements of earth, fire, air, and water that our Greek forebears considered the constituent parts of being [25]. The *Titanic* was obviously the representative of water, and cathedrals were the essence of earth. *Hindenburg*, I took as the symbol of air. Significantly, fire fitted into each story as an agent of destruction involved in all of these transports. For, as we shall see, fire actually played a potentially intriguing role in the *Titanic's* demise. While fire and the destruction of the *Hindenburg* are synonymous, several Gothic Cathedrals have been ravaged and then virtually destroyed by fire down the centuries [26]. Thus the materials and motive energies of these transports are all vulnerable to conflagration. Of *Titanic's* 'fire' we shall hear more soon. These cogitations on commonalty were enough to start me off in the directions which now follow. From the heights of technology to its depths of destruction let us consider the individual stories of the three respective transports – our chariots await!

Reference Notes: Introduction

[1] Cox, M. (2006). *The meaning of night.* (p. 247), W.W. Norton: New York.

[2] As far as possible, I have tried to provide extensive references to original sources and derived texts on the various historical aspects of the present transports.

[3] http://www.titanichistoricalsociety.org/ See also: http://en.wikipedia.org/wiki/Titanic_%281997_film%29 In 1999, I had the honor to present the Arnold M. Small lecture of the Human Factors and Ergonomics Society in Houston, Texas. The subject of my talk was "Custer and the Titanic; Denominators of disasters." This served to link together why many common elements conspired to end the life of both Brevet Major General George Armstrong Custer and the White Star Line Luxury liner.

[4] Remarque, E.M. (1929). *All quiet on the western front.* (Putnam & C.: London, 1970). It is perhaps no coincidence that *Julien Fellowes'* highly successful BBC series 'Downtown Abbey,' actually begins with Titanic's sinking and its greater effects on society and its more specific effects on the family via the "entail."

[5] The Quote is from Robert Browning: "Pippa's Song"

> *The year's at the spring,*
> *And day's at the morn;*
> *Morning's at seven;*
> *The hill-side's dew-pearled;*
> *The lark's on the wing;*
> *The snail's on the thorn;*
> *God's in His heaven —*
> *All's right with the world!*

[6] The United States lost just over 58,000 troops in the official numbers for the whole Viet Nam conflict, the British lost just over 57,000 troops either killed (nearly 20,000) or injured on that one day; July, 1st, 1916.

With the damage to the French troops, the numbers proved greater for one day than for the whole Vietnam conflict, although all numbers of war are very difficult to substantiate. And see: Cave, N. (1999). *Somme: Delville Wood* Pen & Sword: Barnsley, Yorks; England.

[7] What had been an absurd and disastrous mistake at Balaclava which was then turned into an heroic narrative by poets such as Tennyson and the military through agencies such as the founding of the "Victorian Cross" turned out on the fields of France to be an exercise in utter human futility.

[8] Hancock, P.A. (2009). *Mind, machine, and morality.* Ashgate Publishing, Aldershot, England.

[9] Lord, W. (1955). *A night to remember*. R & W. Holt: New York.

[10] see e.g., Miller, M. (1997). *Chartres Cathedral*. Riverside Book Co.: New York.

[11] See compare and contrast: Friedman, T.L. (2005). *The world is flat: A brief history of the twenty-first century*. New York: Farrar, Straus and Giroux. And also: Fukuyama, F. (1992). *The end of history and the last man*. New York: Free Press.

[12] The quotation is from William Wordsworth's "Composed Upon Westminster Bridge, September 3rd, 1802." It is relevant here, not for the buildings seen, but for that momentary intake of breath when wonder strikes as clearly it did for Wordsworth.

> *Earth has not anything to show more fair,*
> *Dull would he be of soul who could pass by*
> *A sight so touching in its majesty*
> *The city now, doth like a garment, wear*
> *The beauty of the morning; silent, bare,*
> *Ships, towers, domes, theatres and temples lie*
> *Open unto the fields, and to the sky;*

> *All bright and glittering in the smokeless air.*
> *Never did sun more beautifully steep*
> *In his first splendour, valley, rock, or hill;*
> *Ne'er saw I, never felt, a calm so deep!*
> *The river glideth at his own sweet will:*
> *Dear God! the very houses seem asleep;*
> *And all that mighty heart is lying still!*

[13] This effect is reminiscent of Galinsky's visual 'foreshortening' which John Senders has brought to my attention. And see: Galinsky, A. (1951). Perceived size and distance in visual space. *Psychological Review, 58*, 460-482.

[14] And see: http://en.shafaqna.com/other-services/other-religions/item/23885-stone-liners-plying-the-oceans-of-faith.html. And also Stephen Platten's description of Lincoln Cathedral in the following terms: *"Lincoln has perhaps the most stunning setting of any Cathedral in England; it floats like an ocean liner on the cliff that divides Lincolnshire in two."*

[15] Miller, M. (1996) *Chartres Cathedral*: Pitkin, England.

[16] The word here is chosen carefully; it is absolute "revelation." The centrality of revelation in human experience derives from my own cognitive anthropic principle which specifies why we humans must necessarily occupy a position in the Universe in which all is not random, but neither is it deterministically perceivable.

[17] see e.g., Eschapasse, M. (1973). *Reims Cathedral*. Caisse Nationale des Monuments Historiques: Paris.

[18] See: http://en.wikipedia.org/wiki/Reims_Cathedral.

[19] I am certainly not the only individual to have felt and identified these curious emotions. Consider for example, *"In France, we also went to the Cathedral in Rhiems, cool in the summer and warm in the winter because*

of how thick the walls were; with all the allied bombing, it was left standing intact. And it was at Rhiems that Germany surrendered unconditionally in 1945." Knight, B. (2002). *Knight: My Story.* Thomas Dunne Books: New York.

[20] See: http://en.wikipedia.org/wiki/15_minutes_of_fame

[21] Protagorans, in opining the "man is the measure of all things" is at once both spectacularly right on a human scale and spectacularly wrong on an absolute scale.

[22] Penn-Jones, Jr, W. (1967). *Forgive my grief.* Midlothian Mirror Inc,: Midlothian, TX.

[23] Gregory, M. (2004). *Diana: The last days.* Virgin Publishing: London.

[24] Sullivan R. (2012). *Untouchable: The strange life and tragic death of Michael Jackson.* Grove Press: New York.

[25] see: http://en.wikipedia.org/wiki/Classical_element

[26] See: http://en.wikipedia.org/wiki/Coventry_Cathedral

"The fighting Temeraire,
Built of a thousand trees." [1]

2. GHOSTS OF THE TEMERAIRE

2.1. Phantoms on the Quarterdeck

In order to frame each of the individual stories of the respective transports that I focus on, I first want to found my discourse on one wonderfully evocative, and indeed provocative, vision. Sadly, this vision is not my own. Rather, it derives from the inspiration of a nineteenth century artist who created an image that has become one of the most emblematic of a nation. The artist was Joseph Turner; his vision was the '*Temeraire,*' Figure 2-1.

Figure 2-1. *My Darling — Joseph Mallord William Turner (1775-1851). (Photograph of the image by the Author). The formal name of the picture is: "The Fighting Temeraire tugged to her last Berth to be broken up, 1838."*

© Springer International Publishing AG 2017
P. Hancock, *Transports of Delight*, DOI 10.1007/978-3-319-55248-4_2

I have to say that from my perspective, Turner's "*The Fighting Temeraire*" is arguably one of the more inspiring, if not the most evocative, images ever created [2]. It normally hangs in Britain's '*National Gallery*' which is situated on one side of London's famous Trafalgar Square. Here, within the Square, and significantly as it turns out, proudly stands Nelson's Column. The painting is clearly one of the gallery's most popular and prized exhibits and has been voted by one poll as Britain's greatest ever painting [3]. Its notoriety is, of course, linked to its unforgettable imagery but as the great American satirist, Damon Runyon would say; "*a story goes with it*," and what a story it is. It is the various levels of this overall narrative that makes the "*Fighting Temeraire*" an iconic representation for my whole book and why I am starting with it.

2.2. A Personal Perusal

For some time then the *Temeraire* [4] had been one of my very favorite images. I intended that, if I was ever presented with the chance, I would see it in situ. When that opportunity came, I was intrigued to see that upon closer inspection there were what appeared to be shadowy and barely distinguishable images of what might perhaps be people standing on the quarterdeck of the *Temeraire*. Were those figures actually there, or was it just my imagination that suggested that Turner had put in such images perhaps as ghosts of those individuals who had fought on board this legendary vessel? I asked about this possibility at the gallery's inquiry desk and was both surprised and gratified to learn that a whole text had been written about this one painting [5]. In this latter book, I learned that Turner had taken several liberties with reality (but then what artists do not?). The presence of the sunset was Turner's imposition and the *Temeraire's* masts would not have been present during her last journey. The ship was actually known by her own crew as the 'saucy' *Temeraire*. Thus Turner's reference to the *'fighting Temeraire'* is, to a degree, his own and adds to the poignancy of his vision. Despite the many years that still remained in Turner's career after creating this painting, the critic John Ruskin called this image *"Turner's last great painting."* Turner himself was so attached to it that he refused to sell it and even referred to the painting as *"my darling."* Upon his death, Turner left the painting to the nation. Even today I have still yet to ascertain for certain whether there are indeed figures on the ship itself. However, I very much like to believe that there are. Perhaps this part of my story was even whispered to me by one of these *"phantoms of the quarterdeck."* Whether there truly are people there or not, the painting itself is a wonderful achievement, and that unique success is not confined to one dimension alone.

There are at least three clear and manifest levels of analysis with respect to Turner's image. Let me lay them out. First, there is Turner's immediate image. In this, we can see the mighty ship of the line, one of the last remnants of the great age of sail, as she is ignominiously hauled up to her destruction by some little, small, blackened 'kettle pot' of a tug redolent of the 'new age' of steam [6]. It is the ascendency here of new technology over old that we are asked to witness. Like Turner, we surely remain unconvinced that this new version of technology is necessarily better than that which has preceded it. Turner forces us to ask if this degenerative superposition; that is the nominal 'bad' driving out the 'good' of technology, is always true. For instance, ask yourself here how much you like speaking with automated phone technologies versus a personal human contact?

It is the final indignity for the *Temeraire* a '*mighty heart of oak*' to find herself helpless under the control of, and subjugated to, the unwanted ministrations of this squat, inelegant and utilitarian little upstart. This is true especially because in her history *Temeraire* had battled successfully against the mightiest enemy warships on the high seas. It is this juxtaposition and our knowledge that this mighty warship is now being dragged to her doom by some little 'kettle-pot' up to some anonymous ship's 'knackers-yard' [7] that rightly stirs our emotion. Like the demise of anything beautiful and courageous there is an inherent tragedy, and Turner has captured and pictured it here.

2.3. The *Temeraire* at Trafalgar

But there are further levels of insight, indeed many more. It is not only the elegant age of sail which is dying. Most probably however, this is why Turner introduces the factually incorrect but artistically justified appearance of the setting sun as a visual recapitulation of the end of an age; thus dies the remorseful day [8]. However, the *Temeraire* had a personal history well beyond being the icon for the era of wind-based sail technology that she here epitomizes [9]. To understand this fully we need to know the context of her greatest triumph and, in reality, *Temeraire's* only major fleet engagement. This is best viewed, metaphorically speaking, from the decks of another ship we can still see and still physically stand upon ourselves. For today, if we want to, we can go down to the Naval dockyards in Portsmouth, England and walk upon the decks of H.M.S. *Victory*. The *Victory* was the renowned flagship of Admiral Lord Horatio Nelson and led one of the two English lines (*Victory* led the *Weather* column, as opposed to the *Lee* column which was led by Admiral Collingwood on board the *Royal Sovereign*) at the fateful *Battle of Trafalgar* on October 21st, 1805 [10].

The English victory at *Trafalgar* blunted the imminent threat of Napoleon Bonaparte in his aspiration to invade and conquer England, one of the last remaining countries in Europe beyond his grasp. It was the defeat of the French and Spanish navies in the cold waters of the Atlantic that denied the brilliant French militarist his opportunity to put all of Europe under his control. The death of Admiral Horatio Lord Nelson, shot on the decks of the *Victory* during the battle, had been a heavy price to pay. However, this successful naval action ensured Albion's safety for a decade at least and in reality for much longer [11]. Today, the *Victory* is the physical embodiment of this pride as much as is Nelson's Column outside the National Gallery (where the *Temeraire* painting is hung), or even Nelson's tomb in St. Paul's Cathedral each of which commemorates Nelson personally. In one of the typically strange concatenations of history, the *Temeraire's* artist, Turner, is himself also interred in St. Paul's mere yards from Nelson's own burial place.

Therefore, the second level at which we have to look into Turner's painted image concerns the story of how the *Temeraire* 'saved' the *Victory*. When the English flagship *Victory* had broken through the combined French and Spanish line across the bows of Villeneuve's *Bucentaur*, she was subsequently locked up with the

French flagship *Redoubtable*. The issue between these two ships, and indeed the balance of the whole battle, now lay in doubt. With the rate of injury and fatalities experienced on *Victory's* decks, there was the very real and distinct possibility that she would be boarded and taken by the French. The English flagship was *in extremis* and desperate for relief. However, second in line behind the *Victory* came the 'saucy' *Temeraire* in this her one and only fleet action. The French had not reckoned on the bravery of Captain Eliab Harvey [12], his crew, and the 98-Gun vessel under them [13]. Although damaged in a skirmish with the 130-gun, four deck *La Santisima Trinidad*, the *Temeraire's* subsequent and very timely broadside disabled the French flagship and Nelson and the crew of the *Victory* were rescued. The French Captain of the *Redoubtable*, Jean-Jacques Etienne Lucas was in the prime position to record *Temeraire's* timely action. His eye-witness account reported that: "*the three decker (Temeraire), who had doubtless perceived that the Victory had ceased fire and would inevitably be taken, ran foul of the Redoubtable to starboard and overwhelmed us with the point blank fire of all her guns. It would be impossible to describe the horrible carnage produced by the murderous broadside of this ship. More than` two-hundred of our brave lads were killed or wounded by it*" [14]. No wonder the *Temeraire* was later referred to as: "*Pride of England and Terror of France.*"

It is no exaggeration to say that it was by the *Temeraire* that the *Victory* was rescued and, indeed, that through the *Temeraire* that victory was won. Following this pivotal moment of the battle, the English subsequently swept the French and Spanish navies from the seas. Although badly damaged by a series of encounters, *Temeraire* still went on to capture two French ships which further enhanced her public reputation and renown. However, this reputation was perhaps most helped by the fact that Captain Harvey was able to communicate directly with Admiral Collingwood just before the latter sent his earliest news dispatches of the battle back to England. As a result, *Temeraire* stood high in the consciousness of the nation and her name became inextricably linked with glory and victory. Alongside the *Battle of Waterloo*, for which the *Battle of Trafalgar* was the sea-born prelude, *Temeraire*'s action represents one of the very highest water-marks of British military achievement and a peak moment of the birth of the later dominant British Empire [15].

For one hundred years Britain remained unchallenged on the seas and unconcerned on land [16]. This was the dominion that the 'saucy' *Temeraire* had bought with her bravery. She was truly one of Albion's heroes. But like all military heroes, when the guns go silent and the flags are struck, they can become something of an embarrassment. It is more than ironic then, that when the Admiralty decided to retire *Temeraire*, she was formally under the command of Captain Thomas Fortescue Kennedy. By coincidence, it had been Kennedy himself who was *Temeraire's* first lieutenant at *Trafalgar* and who had led the boarding party on to the French ship *Fougeux* to secure *Temeraire's* ultimate victory. He must have been saddened indeed to have had to condemn his old ship to its tragic fate [17].

Figure 2-2: *An illustration of Captain Harvey of the Temeraire fighting off the enemy. The illustrated narrative is derived from Collingwood's distorted early dispatches on the course of the Battle.*

2.4. Transports of Delight

So, the story behind the *Temeraire's* image is not simply the demise of the age of sail. No! The backstory here is the sad and shameful way in which this grand old lady of the fleet, and to whom the very existence of the nation was perhaps owed, is being smuggled up in a surreptitious, 'hole and corner' manner, to Rotherhithe to meet her ignominious end. Turner here is implicitly asking us; is this any way for a nation to treat its heroes [18]? This is, in its essence, one of the main foci of my narrative. For here, I will look at the heroic side of technology. It is about such technologies that we can ask so many questions. Can indeed a technology be heroic? How and when do such technologies, grow, blossom, burgeon, and then almost as suddenly collapse dissolve, and disappear from the 'stage' of life? Are we humans wedded, self-symbiotically, with our technologies? Or are we actually becoming, as Henry David Thoreau suggested, simply the *"tools of our tools?"* It is these questions that form the foundation and framework for my exposition.

The third and final level of Turner's image, as discussed here, is represented by the sad and lonely demise of each and every one of us, animate and inanimate alike. Unlike the manner of the *Temeraire's* silent and essentially anonymous departure, I look here particularly at technologies that have literally gone out with a bang rather than a whimper [19]. Although I discuss the genesis, evolution, and demise of these specific technologies, there hovers a larger story behind even these general propositions. For each technology sub-serves human needs, wishes, desires, and dreams. And so, my final aim is to elucidate the ways in which such technologies serve to fulfill these human goals. To examine how such 'vessels' take us to the places that we wish to go to, and in the manner we wish to get there; each by providing *"Transports of Delight."* To begin these stories; I start with the story which is closest to us in the present. Subsequently, I work my way back in time to successively older examples. Eventually I will look here to reverse this temporal trend in order to extrapolate from the principles I establish, to provide an understanding of the technological future for us all.

Reference Notes: Ghosts of the Temeraire

[1] From: The Temeraire by Herman Melville.

[2] The official name of the painting is actually: *"The Fighting Temeraire tugged to her last Berth to be broken up, 1838."*

[3] The poll was conducted by the British Broadcasting Corporation's (BBC) Channel 4 in association with the National Gallery. See: "Turner's Fighting Temeraire sinks the opposition." Nigel Reynolds, Arts Correspondent, 12:01AM BST 06 Sep 2005. In fact, the poll included any and all paintings in the National Gallery which holds such classics as Van Gogh's "Sunflowers." Thus, technically, the approval of the populous went beyond English paintings alone.
http://www.telegraph.co.uk/news/uknews/1497703/Turners-Fighting-Temeraire-sinks-the-opposition.html

[4] A word of French origin, *Temeraire* is often translated as *reckless* or *rash*, but it can also mean *bold* and *brave* or even *audacious* or *fearless*.

[5] see Egerton, J. (1995). *Turner: The fighting Temeraire*. National Gallery: London. And see also: Humphries, P. (1995). *On the trail of Turner*. Cadw: Cardiff; as well as: Bockemuhl, M. (2010) *Turner*. Taschen: Hong Kong. See also: Willis, S. (2009). *The fighting Temeraire: Legend of Trafalgar*. Quercus: London.

[6] Beatson who won the contract to break up the Temeraire for 5,530 Pounds hired two tugs from the *Thames Steam Towing Co.* for this purpose.

[7] Of course this was not an anonymous location. The historical record tells us that John Beatson, who bought the *Temeraire* for scrap, had her towed up from Sheerness to his own Wharf at Rotherhithe, in the Limehouse area of the Thames. In Willis's (2009) wonderful text, he observes that the antithesis of the places of destruction, that is the places of creation, resemble nothing so much as the buidling sites of Gothic cathedrals when he notes: *"The Royal Dockyards were also the focus of the largest skilled communities in the country. Nowhere else could one find such a variety of trades as that which united to build these warships. They were then some of the most sophisticated single structures created by man,*

representing an achievement which has been compared by some historians with the construction of the medieval cathedral." Willis, S. (2009). *The fighting Temeraire.* (p. 99), Quercus: London.

[8] One cannot help but be reminded of Houseman's evocative "May" See: www.chiark.greenend.org.uk/~martinh/poems/housman.html. It ends:

> *"Ensanguining the skies*
> *How heavily it dies*
> *Into the west away;*
> *Past touch and sight and sound*
> *Not further to be found,*
> *How hopeless under ground*
> *Falls the remorseful day."*

[9] *Temeraire* was laid down in 1793 and launched in 1798, being commissioned in March, 1799. She was the second ship of this name in the British Navy. The first had been a French ship captured at the Battle of Lagos in 1759. Like *Titanic, Temeraire* was one of three sisters of the *Neptune* Class. The others were HMS *Dreadnought* and HMS *Neptune.* She was a 98-gun second-rate ship of the line. As the setting sun is emblematic of the demise of sail, so the rising moon is representative of the coming age of steam. What Turner means by the waxing moon and dying sun as technical metaphors is very much open to debate.

[10] see Adkin, M. (2005). *The Trafalgar companion.* Aurum Press: London. And also: Pocock, T. (2005). *Trafalgar: An eyewitness to history.* London. The original plan had the *Temeraire* leading the Weather column but this plan was superseded by events and the task of leading belonged that day to *Victory.* Even this was subject to some further dynamics during the actual attack itself, as was reported in the log of the *Conqueror.*

[11] Wooten, G. (1992). *Waterloo 1815.* Osprey: Oxford.

[12] Most interestingly, Harvey was periodically absent from his ship in order to fulfill another role as Member of Parliament (MP) for Essex.

[13] *http://en.wikipedia.org/wiki/HMS_Victory.*

[14] This was by no means the end of the conflict. *Temeraire* rammed the *Redoubtable* unseating many of her guns and, now lashed to the Frenchman engaged another enemy ship, the *Fougeux* upon her open, starboard side. Eventually, *Temeraire* was lashed to both the French ships and hand-to-hand and ship-to-ship fighting continued for more than three hours. Sandwiched between *Victory* and *Temeraire*, the French ships eventually surrendered, but only after most of their officers had been killed. *Temeraire* was extracted by the *Sirius* whilst still fighting a newly engaged enemy.

[15] While one can argue that the glory of empire was enjoyed by those who followed, see: Morris, J. (1968). *Pax Britannica: The climax of Empire.* Harvest: San Diego, this was the high point in military terms.

[16] And see: Bonney, G. (2006) *The battle of Jutland 1916.* Sutton: Stroud, England.

[17] For more information, especially about Lt. Kennedy, see: Willis, S. (2009). *The fighting Temeraire: Legend of Trafalgar.* Quercus: London.

[18] http://rense.com/general44/dead.htm (Pentagon Keeps US Dead Out Of Sight).

[19] The quote of course is from T.S Eliot's "*The Hollow Men*" which concludes:

This is the way the world ends
This is the way the world ends
This is the way the world ends
Not with a bang but a whimper.

"Here it comes, ladies and gentlemen and what
a sight it is, a thrilling one, a marvelous sight."

3. WHAT A SIGHT IT IS

3.1. Oh the Humanity!

With the above words, the radio commentator Herbert Morrison decorated his live broadcast on WLS, the Chicago-based, Prairie Farmer Station, on the night of Thursday May 6, 1937. His presence was part of a promotion by American Airlines and their link from the city of Chicago to Europe via a transfer to the famous German airship line. As a result, Morrison was at Lakehurst Naval Air Station (NAS) in New Jersey that night to witness the approach and landing of Luftschiff LZ-129 'Hindenburg' on its first Atlantic crossing of the season. Seconds later, his whole demeanor changed. The radio audience heard what remains one of the first and still most evocative and indeed iconic, live eye-witness accounts of a major disaster ever heard [1]. Morrison's commentary still speaks to us across the decades...

"It's burst into flames... get this Charlie, get this, Charlie ... get out of the way please, oh, my, this is terrible, oh, my, get out of the way please! It is burning, burst into flames and is falling on the mooring mast and all the folks, we ... this is one of the worst catastrophes in the world! .. Oh, it's four or five hundred feet into the sky, it's a terrific crash ladies and gentlemen   ... oh, the humanity   ... and all the people!'

© Springer International Publishing AG 2017
P. Hancock, *Transports of Delight*, DOI 10.1007/978-3-319-55248-4_3

Figure 3-1: *The marker showing the site of the Hindenburg Crash against the sky of a New Jersey summer. (Photograph by the Author).*

3.2. DEKKA: Luftschiff *Hindenburg*

It had been three days earlier, on the evening of Monday, May 3, 1937, that the short-wave wireless station in Hamburg, Germany had broadcast a warning of storms over the North Atlantic. The Hamburg call had been addressed to a craft whose call-sign was DEKKA, the short-wave signature of *LZ-129* Luftschiff '*Hindenburg*.' At that time, *Hindenburg* was already outbound from Frankfurt, Germany making for Lakehurst, New Jersey [2] and its fateful and final rendezvous with destiny. Even before take-off, there were signs that things were not as might normally have been expected. *Hindenburg* carried only about half of her possible compliment of passengers but more than one and one-half times her normal crew roster. Interestingly, the crew included a total of five rated airship captains, each of whom could have commanded *Hindenburg*.

Although the company's senior Captain, Ernst Lehmann was on board, the man in command of this trip was Captain Max Pruss. *Hindenburg* also carried three Luftwaffe Officers, nominally as observers for long-range navigation. It was the consensus of the regular crew however, that these officers were part of an especially heavy security effort in response to the threats that had been received against the craft. There were many warnings and one such example letter was in the personal possession of Captain Lehmann and thus was almost certainly known to Colonel Fritz Erdmann, the ranking Luftwaffe officer on board [3]. It was from one Kathie Rauch of Milwaukee, Wisconsin to the German Ambassador in the United States. It read:

"You will excuse my bothering you, but I must send you a friendly warning, one which you will please convey further. Please inform the Zeppelin Company in Frankfurt-am-Main that they should open and search all mail before it is put on board prior to every flight of the Zeppelin Hindenburg. The Zeppelin is going to be destroyed by a time bomb during its flight to another country. While I cannot explain the secrecy of this, please believe my words as the truth, so that no one later will have cause for regret. You must appreciate that this friendly warning is no joke, and also stop the flights of the airship if you are to save peoples' lives. Later can I send you a clearer report

With the word of the truth you will be enlightened." [4]

What the respective authorities made of this and other warnings is difficult to establish. However, it is hard not to agree with the *Hindenburg's* crew that the changing geo-political conditions had induced an ever-increasing necessity for enhanced security. Such conditions and concerns are reminiscent of our present day's circumstances concerning contemporary heavier-than-air '*transports.*' [5]

Colonel Erdman, the leader of the Luftwaffe party on board, had himself been responsible for a most unusual incident just before take-off. When all was ready for departure, he had requested, and received permission to delay the take-off while his wife was let back onto the Zeppelin one more time to again say goodbye to her husband [6]. One experienced airship passenger, Leonhard Adelt, remarked that he had never witnessed such an exception in all of his airship travel. Colonel Erdmann later told how a 'terrible feeling' had come over him. He was not the only one to claim to have presentiments and premonitions of the forthcoming catastrophe.

Figure 3-2. Postal cover from the Hindenburg (from the Author's collection).

Just before dawn on Tuesday, May 4th, having danced among storm fronts all night [7], *Hindenburg* was over the Channel Islands and heading out south of Ireland over the gray Atlantic Ocean. Willy Speck, the wireless officer, switched to the Atlantic band of 600 meters and tried to 'establish,' (i.e., contact) Cape Race, the very same radio station that appears in the story of '*Titanic*' that I shall recount later [8]. As a result of the weather, *Hindenburg* continued to make only slow progress throughout that day. However, by days' end, the airship had settled down into its rhythm and routine that had characterized each of her prior trans-Atlantic round-trips of the previous season.

3.3. Leaving Old Worlds

At 7:45 am on the morning of Wednesday 5, May, 1937, Mackay radio relayed weather warnings from three ships at sea to *Hindenburg*. One of the warnings, coincidentally, was from the *Britannic*, the sister ship to the *Titanic* [9]. It told of light rains extending down the eastern seaboard of North America but shortly after this communication, an electrical disturbance caused a wireless blackout for several hours. The *Hindenburg* continued on uneventfully, passing close to *Titanic's* grave site, until late in the afternoon when the passengers could see the lighthouses of Cape Race and white icebergs contrasting against the surrounding gray seas. Breakfast on that fateful Thursday found the *Hindenburg* over Boston making directly for her New Jersey destination.

The light rain on the coast had, in some ways, been a blessing in that it had helped to extinguish some ground fires at Hindenburg's Lakehurst destination that had, by then, blackened over two thousand acres of forest land adjacent to the Naval Air Station itself. However, any such blessing was a mixed one since Lakehurst continued to still be plagued by rains. Its mast was the only close location at which *Hindenburg* could dock. The next available site, many hundreds of miles away, was in Akron, Ohio. Typically, to save on ground crew costs, landings were planned for specific times and since the 6:00 am window had already passed, most of the Lakehurst professional personnel had now become aware that the giant airship would not be landing, at least until much later that afternoon [10].

3.4. Reaching New Worlds

This change of schedule was not immediately known to the passengers' friends, many of whom had been at Lakehurst since before dawn that day. The anticipated time of arrival at 6:00 am was now well passed. Through the offices of the various media, it became clear that *Hindenburg* would not be arriving until later that afternoon [11]. She would then be looking for a record turn-around time in order to put her back on schedule. Passing over Providence, Rhode Island, the passengers were eating their last scheduled meal on board to be followed at 3:00 pm by the sighting of New York and the subsequent 'pass' down Manhattan in close proximity to the Empire State Building. From there the airship went on to traverse Ebbets Field where the Dodgers and Pirates momentarily suspend their play so that spectators and

players alike could view the spectacular sight [12]. Quickly, the
Hindenburg progressed on, turning out over the Statue of Liberty
toward Lakehurst and her fateful rendezvous.

Figure 3-3. *Hindenburg passing over the skyscrapers of New York*

After asking for a report on local weather at Lakehurst,
Captain Pruss commented that the crew's dinner would have to
be served on board in order to facilitate the anticipated fast turn-
around. At just about 4:00 pm Lakehurst (call sign NEL) sent to
Hindenburg, "*gusts now to 25 knots*" [13]. This wind level prevented
landing and caused *Hindenburg's* decision to turn over Lakehurst
itself and move off out toward the Jersey coast. Captain Pruss
dropped a written message as he crossed the airfield. It read:

"*Riding out the storm.*" As a consequence the spectators and ground crew at the Naval Air Station again experienced a sense of anti-climax as the airship then proceeded to disappear from view.

Minutes later *Hindenburg* was in sunshine out over the coast. After a long loop she was deciding whether to turn again when a radio message from NEL announced "*Thunderstorm moving out*" and at 5:00 pm the Lakehurst' steam whistle called the ground crew back to the station in confident expectation of landing. However, again, all did not run smoothly. Several messages were exchanged and a fresh cloudburst soaked the already dampened ground-crew who had tried to take shelter around the landing mast. At 6:23 pm Commander Rosendahl, in charge of ground operations, radioed *Hindenburg*, "*recommend landing now*," to which Pruss replied "*course, Lakehurst.*"

3.5. Landing the Airship

All was not normal aboard the airship either. At about 7:00 pm, Chief Rigger Ludwig Knorr, who would not survive the coming conflagration, noticed that Gas Cell No. 4, located at the juncture of the body of the airship and the enormous tail assembly, was under-inflated. It was somewhat unusual to find one of these gas cells suddenly becoming 'light' and he mentioned his observation to Chief Steward Heinrich Kubis, who did subsequently survive the crash. Kubis in turn began to re-organize his own plans knowing that if the leak were serious it would certainly delay the hoped for fast turn-around. At 7:10 pm *Hindenburg* received the following message: "*conditions definitely improved recommend earliest*

possible landing" at which Captain Pruss sounded landing stations
and began his approach. On seeing the airfield, Radio Officer
Speck sent one last message to Germany before the trailing
antenna was reeled in. It read "*ready for landing, bad weather.*"
Misinterpreted at the other end, Radio Quickborn erroneously
advised Frankfurt "*Hindenburg landed.*" This news was actually
picked up by *Hindenburg's* sister ship *Graf Zeppelin*, at that time
herself making homeward back across the Atlantic. The reality, of
course, was to prove very different.

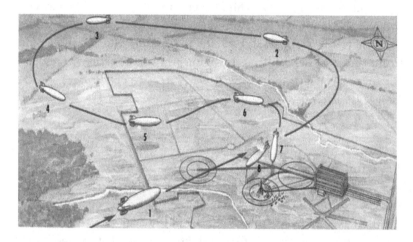

Figure 3-4. *Hindenburg's approach to the landing site at Lakehurst Naval Air Station.*
(Someone else's Figure, need permission to reproduce)

    As the *Hindenburg* came in at 600 ft over the south fence of
the airfield, Herb Morrison began his now famous speech. Captain
Pruss made one pass over the airfield to establish his exact altitude
and to balance the ship that had altered its trim due to the rain's
saturation of the outer skin. Engines were put to idle and gas was
valved in order to bring the ship down toward the landing

position. Trimming was achieved by release of water from holding tanks and at 7:21 pm, at 200 ft. starboard and port handling lines were dropped from the nose of the ship [14]. Several observers saw the ropes smack against the ground and raise a 'cloud of dust' indicating that the ropes themselves were dry. These ropes were quickly attached to capstans to begin to winch the mighty airship down to the ground.

3.6. Disaster!

It was at 7:24 pm when the trouble started. Lieutenant R.W. Antrim, who was himself booked on *Hindenburg's* return passage that evening, saw from his vantage point at the top of the mooring mast what looked like fabric "*very loose and fluttering*" behind the port engine. On the ground news correspondent Alice Hager also saw this 'rippling' effect in the skin of the airship. Under the airship itself, Fred Tobin noticed the asymmetric position of the upper and lower rudders and Francis Hyland, a line handler saw something he took to be an engine backfire. This was perhaps the same thing seen by a Mr. Groves who thought it looked something "*like static electricity*." The official in charge of ground operations, Commander Rosendahl an experienced airshipman himself, saw a "*small burst of flame*" forward of the vertical fin which he knew immediately spelled the end for the vast airship.

Inside the airship, two crewmen in that same lower fin observed what they believed to be the triggering event in and around Gas Cell Number Four. Chief Engineer Sauter reported seeing something like a flash bulb discharge with a well-defined location toward the center of that cell [14]. This was very near the point at which the main catwalk passed through it. Helsman Lau observed something very similar and recognized that the focal point was very close to the maneuvering valves located in that area. Mechanic Richard Kollner heard, rather than saw, the event which he described as a 'popping' sound. Rigger Freund also did not see the triggering event directly but later recalled a muffled detonation and then being engulfed in flames. [15]

Figure 3-5. One of the most famous images of disaster ever captured. The aft-section of the Hindenburg explodes behind the mooring mast. Many details of the disaster are evident in the picture.

Outside of the ship, others were now seeing the famous sequence that is impressed upon our collective visual memory as an icon of these times. One of the most articulate of the first-hand commentaries on the explosion was provided by Mrs. W.R van Meter who observed that "*the stern lighted up sort of like a Japanese lantern … flames inside, swirling around and lighted up the covering of the bag so that I could see the framework of the ship through the covering"* [16]. She was joined in her judgment by many others with their comparable perceptions of the explosive dynamics. However, since a picture is worth many thousand words, the illustration in Figure 3-5 shows the conflagration in its full fury. Commentator Morrison compared the explosion to *"a million magnesium flares"* and numerous motion and still photographers captured the stunning events of the next 32 seconds.

Figure 3-6. *Illustration of the final moments of the Hindenburg.*

Many of those on the ground thought that the explosion and the subsequent crash were simply unsurvivable. In a state of traumatized shock, and trying to reconcile themselves to their loss, many family members even left the airfield convinced their loved ones were already gone. Inside the airship, people were struggling to survive. Many of those observing from the ground were aware of the *Hindenburg's* fate before most of the passengers and crew. Inside, those who could not see the explosion directly were aware of a 'shaking' and some individuals actually first became aware of the extent of the catastrophe by directly observing the reactions of those on the ground below them.

For a moment, there was a *"remarkable stillness"* and then - explosion. Seconds later, the stern began to sink and passengers and crew were thrown on top of each other as the ship lost all stability. In all parts of the airship, those on-board now realized the scale of the disaster. What is so miraculous about the crash of the *Hindenburg* however, was the number of survivors. Indeed, survivors far exceeded the number of the dead (35 fatalities out of 97 on board) [17]. However, viewing the contemporary newsreels of the crash, it is hard not to agree with those less sanguine of the live spectators, that all would be lost. Remarkable tales were told by those who miraculously escaped. For example, all five rated Captains made it from the Zeppelin alive! Only Captain Lehmann subsequently succumbed to his injuries. The two individuals who actually saw the explosion from inside, Rudolph Sauter and Helmut Lau, also both survived despite their proximity to the explosion itself. Some of the younger passengers like Peter Belin succeeded in jumping to safety. He was alongside the entertainer and acrobat Joseph Spah who after 'bouncing' off the ground picked himself up, brushed himself off, and went to find respite with his waiting family.

Perhaps the most remarkable story of survival is the wonderful departure made by Mrs. Marie Kleeman. In the course of the crash, the supports holding the regular stairs that were used as the normal exit from the *Hindenburg*, burned from their fittings. Under gravity the stairs then angled down and touched the ground to provide an available escape route. It was in this fashion that Mrs. Kleeman made her bizarrely 'normal' departure in most abnormal of all circumstances [18]. Many of the crew fared almost equally as well. Jonny Doerflein and the cabin boy Werner Franz were both showered by the punctured ship's water tank and consequently, like several others, emerged with few if any burns or injuries. As with all stories of disaster, there were heroes and there were villains. Captain Pruss bravely returned to the conflagration in a sadly unsuccessful attempt to rescue wireless operator Willy Speck. Others of the American landing crew fought fear and fire to help fleeing survivors. As Hoehling so evocatively observed:

"The Hindenburg, a twisted, hotly smoldering skeleton lay near the muddy obscure Paint Branch Creek, 4381 miles from Frankfurt, 77 hours in time. Here in tortured death reposed the alpha and omega of the super-airship, the dream of giant luxury liners of the sky. The era of the Zeppelin had passed in 32 seconds in the wet New Jersey twilight. All at once, the rigid hydrogen filled airship was extinct. It had passed even as the dinosaur. [19]."

Figure 3-7: *The wreckage of the Hindenburg lies 'beached' upon the Lakehurst landing ground. It was not the only aviation disaster of that summer. Less than two months later on July 2nd, Amelia Earhart disappeared in the South Pacific and was never heard from again.*

3.7. The Crash and Its Aftermath

The living story of the Zeppelin *Hindenburg* was now over but its mystery still persists to this day. The wreckage was divided up and removed from the precincts of the Naval Air Station. Today, only a small memorial to her demise remains, (see Figure 3-8). With her went the last of the great airships as both the *Graf Zeppelin* and the *Hindenburg's* own partially completed sister ship *LZ-130* were decommissioned almost immediately afterward. *Hindenburg* was the *'Cathedral of the Skies'* that crashed and burned that summer evening now almost eighty years ago. The technology itself became essentially moribund with this one failure and has still sadly never been effectively resurrected to this day. Yet perhaps the most interesting mysteries of the *Hindenburg* remains quite simply why did she crash? Theories abound and here I consider only some of the major candidate explanations [20].

Figure 3-8. *A picture of the Hindenburg Crash Site today with the Lakehurst Naval Air Station Hanger's in the background. (Photograph by the Author).*

3.8. Why *Hindenburg* Crashed

The immediate response to the crash was "sabotage." *Hindenburg* had been threatened specifically and the presence of the Luftwaffe officers aboard encouraged this belief of a direct, and now delivered terrorist act. Sabotage theories ranged from the incredible, such as phosphorescent bullets being fired from the woods surrounding Lakehurst, to the eminently feasible, such as a small incendiary device placed close to, or even within, Gas Cell Number Four; assuming some source of oxygen could have also reached the ignition site. Accounts given by those closest to the purported point of ignition tended to support this later interpretation.

Immediately following the crash, there were two official inquiries set up into the disaster [21]. They came at the matter both literally and figuratively from very different directions. The German Inquiry, under Dr. Hugo Eckner, certainly sought to exercise damage control and its members were anxious that any 'cause' was not associated with the vehicle itself. In this way, the 'static' electricity discharge notion soon gained much traction and became their favorite causal candidate. In this version of events, *Hindenburg* was doomed by static electricity, or even St. Elmo's fire, associated with the almost unique atmospheric conditions as she approached Lakehurst that day. Although somewhat unlikely, this "*act of God*" explanation found much favor since it exonerated virtually all the human agencies associated with the crash. Over the years, this explanation has been identified most frequently as the cause of events that day in 1937.

More recently, we have been treated to alternative explanations founded upon a variety of factors such as the composition of the material of the outer skin used to cover the Zeppelin. The latter view is championed for example, by NASA scientist Addison Bain and his proposal is that the events which led to the destruction of the *Hindenburg* first started as a discharge occurring across the fabric skin of the airship and then through its frame. Given the volatility of the composition of the cover and this origin of ignition, it is suggested that the combination of these two major contributory factors proved fatal. No one disputes that, in any and all of the causal accounts, the exploding hydrogen became the primary conduit of *Hindenburg's* destruction. It is simply that Bain's conception focuses upon the materials used to construct its outer cover. Why this had never before proved a particular problem to the Luftschiffbau and her airships remains undetermined, although the proposition protests that *Hindenburg's* cover was, at that time, a relatively 'new' material. It has been claimed that confirmation of this theory was actually evident in specific German tests made in the two months that followed the *Hindenburg* crash.

Upon this basis, it has been suggested that design changes were subsequently made to *Hindenburg's* sister ship, *LZ-130*. However, by then, in the minds of the paying public the damage had been done. This latter theory concerning the unstable nature of the airship's covering is technically termed a *sufficient* explanation for what could have occurred that evening at Lakehurst. What we cannot as yet determine is if it is an *exclusive* account. That is, we cannot say with certainty whether this was either the primary or indeed the sole reason why the disaster occurred. Therefore other accounts which may not yet have been thought of, may still represent the nominal 'true' cause i.e., ground-truth. Bain's theory is a possibility but does not then exclude other interpretations of the source of the hydrogen's ignition. For a number of reasons, especially those involving the politics of the times, I myself think that there was a more intentionally human causal agency involved in the origin of the conflagration [22]. After all, it should be noted that time-bombs had previously been discovered on other airships, including one on *Graf Zeppelin*. And wasn't it oh so convenient that *Hindenburg* exploded right in front of those cameras?

3.9. The End of an Era

So came the end of the odyssey of our first *"transport of delight."* Now all that is left of the stately airship and its peers are flickering black and white images and the enthusiastic eulogies of their impressed passengers. In our world today, the noisy, high altitude, cattle-packed commercial jets are almost the very antithesis of the slow, quiet, stately and even regal, low-altitude progress of the great airships. Hopefully, in some conceivable future, these latter capabilities will be resurrected [23]. Although this reality may be rather far off, surely it will not be long before some surrogate virtual experience is provided. This, at least to the extent that we can all make such an experience 'personal.' In leaving *Hindenburg's* story, let us wish good fortune to those organizations that are working to resurrect the rigid airship. I myself can hardly wait to be so 'transported.'

So now we need to move on from the demise of the *'Cathedral of the Air'* and step to the classic *'Cathedral of the Waters.'* It is a story that, because of several major motion pictures, we collectively now know quite intimately. However, this story bears repeating, especially in light of perpetuated mis-information [24]. So let us slip back in our minds now even further in time to a moment some quarter of a century earlier than the *Hindenburg's* fateful flight. In so doing, we reach back, out, across the dark North Atlantic waters and back in time to the ever-famous *"Night to Remember"* [25].

Figure 3-9. *The memorial stone at the site of the Hindenburg crash today. (Photograph by the Author).*

❖❖❖

Reference Notes: What a Sight It Is

[1]. Morrison's commentary and the newsreel coverage of the disaster can still be heard today, accessed from any number of websites, e.g., *http://www.archive.org/details/hindenberg_explodes*. Indeed, we might well argue that this was the very first disaster captured by modern media. While people had to wait for the newsreel footage to appear, Morrison's voice was "real-time," arguably for the first. It was this 'new' immediacy that lent credibility to Orsan Welles subsequent October 30th, 1938 broadcast of the alien invasion.

[2] See: Archbold, R., (1994). *Hindenburg: An illustrated history*. Madison Press New York. See also: Tanaka, S. (1993). *The disaster of the Hindenburg*. Time-Life: New York.

[3] See; http://facesofthehindenburg.blogspot.com/2009/02/colonel-fritz-erdmann.html.

[4]. Discussion of the letter is given in Hoehling, A. (1962). *Who destroyed the Hindenburg?* Little Brown: Boston.

[5] Russell, P. (2009). *Colonel Fritz Erdmann.* Retrieved from http://facesofthehindenburg.blogspot.com/2009/02/colonel-fritz-erdmann.html

[6] Our current society is now redolent with reports of air attacks of varying sort, post 9/11. This includes actual plots, those reported but necessarily unconfirmed, as well as mysterious disappearances of aircraft such as the Malaysian Airlines MH 370.

[7] Lace, W. (2008). *The Hindenburg disaster of 1937.* (p. 44). New York: Chelsea House.

[8] Thorne, G. (1993). Cape Race: Sentinel to disaster. *The Titanic Commutator, 16* (4), 8-21.

[9] Mills, S. (1996). *HMHS Britannic: The last titan.* Shipping Books Press:

[10] Tony, D. (2010). *Morrison, Herbert; biography.* Retrieved from http://pabook.libraries.psu.edu/palitmap/bios/Morrison__Herbert.html

[11] Lewis, V. (2002). *The Hindenburg.* Retrieved from http://www.vincelewis.net/hindenburg.html.

[12] Snyder, J. (2013). *365 oddball days in dodger history.* Clerisy Press: St Covington, KY.

[13] Rosenburg, J. (n.d.). Hindenburg disaster: The tragedy that ended lighter-than-air passenger travel in rigid dirigibles. Retrieved from http://history1900s.about.com/cs/disasters/a/hindenburgcrash.htm

[14] Kelly, M. (n.d.). *The Hindenburg disaster part 1: The events of may 6th, 1937.* Retrieved from http://americanhistory.about.com/od/hindenburg/a/hindenburg.htm

[15] Russell, P. (2008, December 11). *Faces of the Hindenburg: Rudolph Sauter*. Retrieved from http://facesofthehindenburg.blogspot.com/2008/12/rudolf-sauter.html

[16] Russell, P. (2008, December 10). *Faces of the Hindenburg: Hans Freund*. Retrieved from: http://facesofthehindenburg.blogspot.com/2008/12/hans-freund.html

[17] Hoehling (1962). Op cit., p. 108. In terms of proportional loss of life, the 36.1% of the Hindenburg is much lower than the comparable 68% Figure for the Titanic, as we shall see. It is of passing interest that the survivors of the Titanic at 32% are very near the same percentage of the losses for Hindenburg at 36%.

[18] History.com Staff. (2010). *Hindenburg*. Retrieved from http://www.history.com/topics/hindenburg.

[19] Mara Kleeman leaves the Hindenburg. (p. 120).

[20] Hoehling (1962). Op cit (p. 120).

[21] Young, N, (2013), *What Destroyed the Hindenburg?* [DVD]. Silver Springs MD: Discovery Communications

[22] Eckener, H., Duerr, L., Breithaupt, J., Bock, G., Dieckmann, M., & Hoffman, W. (1937, January 1). German Investigation Commission Report. Retrieved September 15, 2014. Commerce Department Accident Report on the Hindenburg Disaster. (1937, January 1). Retrieved from http://www.airships.net/hindenburg/disaster/commerce-department-report

[23] Cambou, D. (Director) (1996). *The Hindenburg* [Web]. Retrieved from http://whiv.alexanderstreet.com/view/1756265/play/true/

[24] Nichols, D. (2011). The Hindenburg disaster and the end of the airship era. *History Today*, Retrieved from http://www.historytoday.com/dean-nicholas/hindenburg-disaster-and-end-airship-era

[25] Even though I thought I knew the story of *Titanic* well, I have been greatly helped by a recent reference work which provides valuable insight about assumptions and perpetuated myths. See Maltin, T., Aston, E. (2010) *101 Things you thought you knew about Titanic...But Didn't*. Chatham: CPI Mackays.

[26] Lord, W. (1955). *A night to remember*. R & W Holt: London.

"God Himself could not sink this Ship." [1]

4. THE LARGEST MOVING OBJECT EVER BUILT

4.1. Introducing *Titanic*

Almost exactly one-quarter of a century before the *Hindenburg* crash, at 11:40 pm on the night of Sunday 14th, April, 1912, the triple-screwed 46,328 ton, 882 foot White Star Liner *'Titanic,'* the then largest moving object ever built by human hands, hit an iceberg south of Cape Race in the North Atlantic and sank in just over two and a half hours. It was her maiden voyage and she carried a variously reported 2227 to 2235 passengers and crew [2]. Of those individuals, 1522 were lost including her Captain, Edward J. Smith and her designer, Thomas Andrews. It was, and remains, the most famous maritime disaster in all of human history [3]. What made the event even more noteworthy was that the ship itself had been advertised as *'unsinkable'* and thought to be so even by some of these intimately connected with her creation and operation. As epitomized in the headline quotation above, for many people the *Titanic* represented what was then thought to be the final victory of technology over nature [4]. For others such assertions were viewed then, as they still are now, as a direct challenge to God and blasphemous in their very heart. Unsurprisingly, *Titanic's* catastrophic destruction was in consequence, subsequently interpreted in very diverse ways. The fact that she went down on a Sunday was not lost on some.

© Springer International Publishing AG 2017
P. Hancock, *Transports of Delight*, DOI 10.1007/978-3-319-55248-4_4

Figure 4-1. *Titanic at dock*

4.2. Leaving Old Worlds

A coal strike in England had affected all of the trans-Atlantic shipping of that early crossing season. A vessel of *Titanic's* size, which required some 650 tons of coal each day inevitably drained the limited, fleet-wide supply of its parent company, the White Star line. As a result, White Star had to cancel the voyages of both the '*Adriatic*' and the '*Oceanic*' in order to provide enough coal for *Titanic's* maiden crossing. White Star also bought additional coal from elsewhere in the system and these acquisitions disrupted the shipping schedules of several other lines. One result was that many passengers from these other ships were transferred on to *Titanic*. Today, we can still see evidence of this fact in the form of tickets for other vessels with the *Titanic's* transfer designation on them [5].

Figure 4-2. Titanic shown in the process of departure.

Disaster had threatened *Titanic* at the very beginning of her maiden commercial voyage. As she was departing the dock in Southampton, the enormous suction generated by her unprecedented size pulled another liner, the *New York* toward the

Titanic, snapping a retaining hawser. Captain Smith immediately ordered 'port engines' and with the help of tugs, managed to 'push' the *New York* away. Fortunately, or unfortunately (depending on hindsight), *Titanic* was then able to proceed on her way undamaged. On this occasion Captain Smith was nominally luckier than he had been earlier when in command of *Titanic's* sister ship, the *Olympic* (see Figure 6-3) During his tenure as captain of that vessel, the *Olympic* had collided with HMS *Hawke* in the Solent; the channel that connects the port of Southampton with the English Channel [6]. Although 'lucky' in avoiding disaster in these initial moments of departure, as we now know, Smith's luck on *Titanic's* maiden voyage was, spectacularly, to run out.

Titanic had been suffering from a coal bunker fire which might even have started as early as Tuesday 2nd April. Shortly after leaving Southampton on the way to Cherbourg on Wednesday 10th April, the crew was looking to extinguish this fire by 'raking out' the affected coal bunker. This fire was not extinguished until the afternoon of Saturday 13th April. One fireman later reported that coal had to be moved from sections two and three on the starboard side. It has even been speculated that when water came in to these sections following the iceberg collision, the bulkheads in these sections would not hold because the weight of the coal which would have served to support them had been removed. From this observation one could, at admittedly a very long stretch, argue that *Titanic* succumbed because of fire! However, this assertion concerning weight-dependent bulkhead resilience is, from an engineering perspective, most probably incorrect. All this was in the future as *Titanic* called at Cherbourg in France for additional passengers and to transfer mail. Her final port of call was Queenstown in Ireland near to which the last image of the *Titanic* afloat was recorded, see Figure 4-3.

Figure 4-3. One of the last pictures of the Titanic above water, proceeding across the Atlantic from the southern coast of Ireland.

❖ ❖ ❖

4.3. MGY: Radio Call Sign of Titanic

The relatively warm winter that year had 'calved' (released) any number of 'growlers' (relatively small individual icebergs) into the 'southern route' of the Atlantic shipping lanes. Indeed, tracing the antecedents of the iceberg, we actually come back to the unusually high snowfall levels in Greenland and the north some years prior to *Titanic's* construction. This propensity for the presence of icebergs was well known to professional mariners of the North Atlantic crossing. Indeed, there was a long history of collisions with icebergs in the north Atlantic. The '*Arizona*' in 1879, the '*Concordia*' in 1899, and the '*Columbia*' in 1911 had each survived direct bow impacts with icebergs. *Titanic*, with her radio call sign MGY, had received, and was actually plotting reports of, the occurrence of icebergs in her area.   Fourth Officer Boxhall noted reports of icebergs to the north of their track and a communication from the *'Amerika'* also indicating ice to the south of *Titanic's* projected route. Whether all of these reports reached *Titanic's* bridge remains a point of some contention [7].

In respect of specific communications, the *'Athinai'* had explicitly reported ice directly on *Titanic's* track. Warnings were also received from the *'Baltic'* and the *'Californian'* [8]. Apparently, *Titanic* did not slow in response to any of these warnings. Indeed, one of the main questions to arise in the formal inquiries which followed the disaster concerned *Titanic's* speed at the time of impact. This has been variously reported at around 21-22 knots. There remains a persistent suggestion that Captain Smith was being urged by White Star's owner and passenger, J. Bruce Ismay to capture the Blue Riband for the fastest Atlantic crossing. [9] This suggestion has been doubted by many experts, primarily on the grounds that *Titanic* could not, and was not built, to challenge for the speed record [10]. Regardless of this contention, it is the case that the ship was traveling at very close to its full speed at the time of the impact. Apparently, specific discussions concerning slowing for icebergs had not been the topic of any extensive deliberation. However, it is important to note that *Titanic's* response (or lack of it) here, simply represented standard maritime practice at that time. Speed was only curtailed when ice had been specifically spotted by personnel on the ship itself.

4.4. "Iceberg – Dead Ahead"

Close to midnight, relatively few people were still up and around when the *Titanic* did collide with the iceberg. Those who were not yet asleep described the collision itself as a jolt, enough to give someone a 'start' but not something that caused immediate or serious concern. Some 'night-owl' passengers came out from the first class smoking room just in time to see the iceberg scraping down the starboard side of the ship and disappearing astern [11]. With its departure from immediate view the short-lived excitement died down and the passengers returned inside, out of the cold night air.

Fred Fleet, one of the lookouts, had shortly before phoned down a belated warning to the bridge to Officer Murdoch. As the ranking officer on deck, Murdoch had looked to take immediate avoiding action and ordered *first* hard-a-starboard, then hard-a-port, stop engines, and finally full speed astern; although the occurrence of the last of these orders is still a topic of contention and debate. Given the brief time interval between the warning and the subsequent collision, it is more than probable that impact was unavoidable. Perhaps the only option open to Murdoch was exactly how, and in what fashion and attitude, *Titanic* would collide with the iceberg. Unfortunately, it was this very attitude of the ship in meeting collision that proved to be the major factor in her eventual sinking. If, for example, she had hit dead on, the probability is that *Titanic,* although quite severely damaged, would have survived.

Figure 4-4: *Artist's impression of the RMS Titanic in full flow.*

Immediately upon feeling the impact, Captain Smith came to the Bridge from his cabin next door. After requesting a status report, he ordered that the watertight doors of the 'Stone' system be closed, but that order had already been given. Despite these and other remedial measures, *Titanic* was already doomed and with the number of lifeboats she carried, so were many of her passengers and crew. Of course, much speculation revolves around the sequence of sighting, warning, and collision. The 'lookout' station that night was no sinecure. Apparently, as was standard, the lookouts did not have binoculars since it was thought these might actually limit the range of each lookout's perception [12]. Without much in the way of any other technical aids, the two men in the crow's nest were restricted in the range and distance over which they could see and react.

Almost all accounts tell us that the Atlantic that night was, in a state of flat calm. *"Still as a pond"* as someone described it. Indeed, this flat calm may well have played a role in preventing the two lookouts getting a sufficiently early detection of the iceberg. In more normal seas, lookouts might have been able to observe the splash and spray around the base of the iceberg but on that fateful night, this vital cue was essentially absent. Also, it has been suggested that what *Titanic* hit was actually termed a 'blue' iceberg. This form of iceberg is one which having recently turned over did not show the typical white brilliance that would again make it more conspicuous on a dark night. This postulate was forwarded as a result of the testimony of lookout Reginald Lee. However, the consensus seems to be against this "blue iceberg" hypothesis. Nevertheless, various factors combined together with the cold conditions experienced by the exposed look-outs probably made for a later detection than would normally have been the case for this vigilance task [13]. Combine this 'late' detection with a high closing speed and the lethal jaws of a crushing impact were inevitably snapping shut.

4.5. Death Throes of a Dream

Of the actual sinking itself, there are stories of cowardice and stories of heroism, as there always is in any narrative of human disaster. Perhaps the most inspiring of the heroic stories is that of *Titanic's* band. They were heard playing, even as the ship slipped beneath the waves. It remains an enigma as to whether they were playing '*Autumn*' (*Songe d'Autumne*) or '*Nearer My God To Thee.*' While most authorities favor the latter and, while the polemic is involving, it is purely academic. What is neither in dispute, nor in anyway academic, is the undoubted courage of each and every one of those individual band members [14]. It is comforting to believe that they all followed Captain Smith's falsely reported but wonderfully evocative injunction to *"Be British Boys!"* Rather like the very fanciful story of Captain Smith swimming with a baby to a lifeboat before swimming back toward the sinking ship, this phrase is one of the myths and legends of the *Titanic* which makes it still the most studied, discussed and perhaps mythical maritime disaster of all time. A final irony in respect of Captain Smith is evident in his oft-quoted 1907 pronouncement:

> *"When anyone asks me how I can best describe my experience of nearly forty years at sea, I merely say uneventful. I have never been in an accident of any sort worth speaking about. I have seen but one vessel in distress in all my years at sea . . . . I never saw a wreck and have never been wrecked, nor was I ever in any predicament that threatened to end in disaster of any sort."*

But let us not be too hard on Captain Smith; for, we should be aware that in time, nature has a way of revenging herself on all human hubris.

It took approximately two and a half hours for *Titanic* to finally succumb to the inevitable. In a rather curious and perhaps bizarre co-incidence, which might actually be no co-incidence at all, *Titanic* and *Hindenberg* both went down at almost the same rate in terms of tons per second [15]. Some lifeboats which had been sent away early had less than half their capacity filled. Few passengers at that time would take sufficiently seriously the proposition that *Titanic* was already doomed. Even when the situation finally became more generally transparent there was still an understandable reluctance to leave the liner. In any event, it was the first and second class passengers, most especially women and children in these classes, proved relatively immune to the fate that would befall more than fifteen hundred of their companions [16]. However, this pattern of demographic salvation was not one that was actually typical of maritime disasters to that time; in this *Titanic* was an exceptional case.

Death came in many forms that night. Primarily it was exposure to, and drowning in, the freezing North Atlantic Ocean that put an end to the vast majority of individuals. Tantalizing glimpses of hope appeared to both the mariners and passengers alike. The wireless operators had initially sent the Morse-code 'CQD' emergency message and subsequently broadcast the newer 'SOS' message to raise ships in the immediate area. *Carpathia* was the one to make the greatest effort steaming at very close to her full speed to cover the 48 miles between themselves and the survivors. However, *Titanic's* lookouts had spotted lights on the horizon and signal rockets had been launched in the hope of a more immediate rescue. Whether these lights were specifically those of the *Californian* and what response her master Stanley Lord should have made, remains the subject of an unending debate.

It may well have been that some rescue might have been feasible but the subsequent pontifications by Senator Smith and Land Mersey, heads of the respective U.S. and British Inquiries, that all lives might well have been saved are probably more in the realm of hyperbole than reality. Of course, hindsight is always crystal clear, human beings being wonderfully wise after the event. And, of course, we were not there that night. Places in the lifeboats now meant life but now the lifeboats were gone and many hundreds were still struggling to survive. Initially four hours away, *Carpathia* was rushing to the scene but she would arrive too late for most. Eventually reaching the transmitted location to secure survivors, *Carpathia* could only review the hole in the ocean into which *Titanic* had disappeared. The sad and scattered remnants of the great liner lay strewn across the morning seas. Survivors were hoisted aboard and at last in the clear light of day the full extent of the disaster finally became shatteringly clear.

4.6. Those Who Survived

The reaction on either side of the Atlantic was one of stunned disbelief. When the *Titanic* went down, it wasn't only a ship that sank. With it went a way of looking at the world as one of comfortable certainty. The inflated arrogance of the Victorian and new Edwardian era had been punctured by one savage thrust of nature. As I have noted, but something that always bears repeating, the vestige of this attitude was subsequently swept away once and for all in the fields of northern France in the '*war to end all wars,*' which to repeat, sadly did not end war at all. The doubts and uncertainties that were raised still percolate through our modern attitudes today. Our relativistic perspectives float, anchorless, like the flotsam of the *Titanic* itself, in a world now bereft of certainty. For the twentieth century at least, the claim that we had 'conquered nature' only echos, *sotte voce*, down the corridor of time in the most tragic whisper of irony.

> *"Because we were so sure. And even*
> *though its happened it is unbelievable."*

It wasn't only the lost who proved of interest, it was also a listing of the survivors who told so much about those times [17]. The class and gender divide were very much in evidence in accounting for those who lived and those who died. Most interestingly, if we count *Hindenburg's* passengers as all 'First Class' then we find overall that the survival rates of the two disasters, like their destruction rate, also prove to be very similar [18]. As it was primarily women and children of the first and second class who were rescued, it was all the other denominations of passenger and crew that were in great danger. Indeed, the very survival of White Star owner Bruce Ismay was considered reprehensible in and of itself. The ethos of the times very much demand that, like the Captain E.J. Smith and architect Thomas Andrews, he should also have *"gone down with the ship."* Ismay's behavior and survival was very much contrasted with the quintessential Edwardian act of self-sacrifice which had actually occurred less than a month earlier. *Titanic* sank on April, 14th, 1912, but on March 17th, 1912 Captain L.E.G. 'Titus' Oates, had, in one of the most heroic gestures of modern times, knowingly sacrificed himself in order that the three remaining members of Scott's South Pole expedition had a better chance of survival [19]. *"I may be sometime"* is surely a phrase Ismay would have hated to hear!

Now unable to sustain the physical forces visited on her, the sinking *Titanic* cracked her back and sank in two pieces below the flat calm of the Atlantic. The noise of the ship's break up was so loud that even some of the sailor's believed, in error as it proved, that *Titanic's* boilers had exploded. Eventually *Titanic* came to rest on the seabed, a ghost of maritime engineering, waiting for Dr. Robert Ballard and his part-time excursion for its rediscovery.

Since that time the wreck has been almost more visited and abused, at least in appearance, than an entertainment theme park. I must admit then I myself succumbed to the blandishments of the Shopping Channel and bought a 'genuine' and 'certified' piece of *Titanic's* coal! As I write today, I have it before me. She was the wonder of her age, a marvelous topic for film makers of all stripes, but a grave for over 1500 souls. Perhaps this is one wreck that we should now leave in appropriately sepulchral isolation.

4.7. Heroes and Villains

As with all disasters, the subsequent narrative eventually proved equally as important, if not even more important, than the facts themselves. The tale here reflects the great human story-telling tradition of heroes and villains. Let us start, in the time-honored fashion, among the heroes. Of course, to be a hero aboard the *Titanic* one essentially had to be dead and all of the inimitable band members that we have met before, must feature prominently here [20]. What was most evident that night was the aural evidence of their collective adherence to duty. As the engineers labored deep in the bowels of the ship to maintain the lighting and other systems, the band played on. It would be churlish and disrespectful to their sacrifice to suggest anything other than it was extraordinary courage that they displayed that night. But they were not alone.

Some have tried to cast Captain Smith in the role of hero, bravely conveying babies to salvation while eventually swimming back to the doomed vessel and sacrificing himself. Legend and myth are so hard to suppress and, after all, Smith was now dead. The famous James Cameron motion picture even features Thomas Andrews, if not in an heroic role, at least in a sympathetic light as a figure of probity and responsibility. But of the villains of course, as I have already noted, Bruce Ismay stands at the forefront of the queue in taking the collective disapprobation of the world. While many of the men in first class drowned, Ismay had the cursed luck to survive. The truth of Ismay's actions is, of course, far more subtle and nuanced than any stereotyped narrative permits. Ismay had actually helped in the process of evacuation and had finally left in one of the last lifeboats to leave the sinking ship when that boat was not yet filled nor had other passengers immediately present to be loaded. And even at that stage many still believed the ship itself would not go down and so a brief 'boating' trip on the freezing Atlantic held no great appeal. In some ways Ismay might have been considered remiss if he did not encourage others to leave the ship through the example of his own actions. However, such qualifying comments rarely if ever fit with what we collectively want to be true and so the mustachioed Ismay was doomed to assume his villainous role when, with only the small matter of his death, he could have been a shining hero!

Nor were the passengers and crew of *Titanic* alone cast into being shoehorned into these polarized roles. While Captain Rostrom of the *Carpathia* was generally acclaimed a hero for doing his duty, Captain Lord of the *Californian* was eventually hounded out of existence itself for what were almost certainly technical shortfalls in maritime technology and/or the specific circumstances of that unforgettable night [21]. The media wanted and needed these actors for their simplistic passion play. After all, the primary *mise en scene*, the ship itself, now lay some 10,000 ft.

down, at the bottom of the Atlantic. She would not make her reappearance on the world's stage for many a decade [22]. It was those individuals left on stage who were constrained now to fulfill their roles at the respective Inquiries in the United States and Great Britain. These hearings, and the associated media coverage, was where these first formal acts of praise and blame were to be officially played out.

The number of theories as to the reason or reasons why *Titanic* sank are now approaching the number of individuals lost that night. They range from the material form of the ship in terms of construction materials and methods, to the control and guidance of the ship itself. Theories feature culpable activities and inactivities of any and all associated individuals. These respective identifications are typical of all such inquiries and their deliberations. Further, when the official proceedings have concluded then the comments and observations of individuals, informed or otherwise, inevitably surface. Finally, the disaster can be accorded the ultimate "accolade" of a Hollywood blockbuster which is either "*inspired by the true events*," "*based on true opinions*," or "*marginally aware that something similar might once have happened*." Hollywood's dictat that the facts should never be allowed to get in the way of a good story is especially evident in the story of "*Titanic*."

The many films that have been made about *Titanic* range from the appealing accurate to the appalling absurd. In the end it is often the myths and fabrications that entrench themselves into the public consciousness until the story of what people think occurred can be in direct contrast with what we can establish as the "ground-truth" of historical events [23]. One obvious recommendation that we can give to anyone involved in any such disaster is to get your "story" out there first to the receptive media and ensure that its narrative is simplistic, dichotomous, and the one that features you as the hero. To achieve the ultimate adoption of your version is contingent upon making it as straightforward as is consistent with credulity. It should present heroes and villains, and again be sure to make yourself the principal hero. This has worked well in the past and, as most individuals have neither the time nor the inclination to explore the full detail of any story, you can anticipate that your version will eventually become received wisdom and the social consensus.

4.8. Summary and Conclusion

Titanic remains the absolute epitome of fortune's reversal. One minute the quintessential representation of ultimate luxury and engineering attainment, the next, literally the sinking symbol of hubris. If one is a deist, the iceberg was God's instrument in punishing the sin of pride. If one were a pure secular humanist the iceberg is simply a tragic but necessary part of nature. If one were an unmitigated ego-centric Manichean, the iceberg was an agent of evil placed by the devil in the path of innocence. None of these perspectives individually capture the full portrait of the disaster. However, the ship that could not sink, the transport that was the last word in opulence, had been wiped from the seas apparently by chance alone. The transport of delight was itself transposed into a legend of tragedy. This reversal happened all in a matter of hours. The story is a salutary lesson to us all that when we reach for the very highest in existence we risk an especially public and tragic exhibition of our shortcomings. It is, in its way, a theistic and indeed moralistic tale. With this firmly in mind, we can now move to more directly spiritual transport with the story which follows.

Reference Notes: The Largest Moving Object Ever Built

[1] It is suggested that this phrase was yet one more of the myths that surround the great ship, among so many. http://www.titanic-whitestarships.com/TandOWSS%20FAQ.htm#Organisations.

[2] see Maltin, T. (2010). *101 things you thought you knew about the Titanic – but didn't.* Penguin: New York, and see also: for example: http://historyonthenet.com/Titanic/passengers.htm

[3] It is not now, and was not then, the largest loss of life in a marine disaster. The protestation is that the *Sultana* held more individuals and had a greater loss of life, especially because many of the latter's passengers were injured soldiers. (See: Huffman, A. (2010). *Sultana: Surviving Civil War, prison, and the worst maritime disaster in American history.* Harper-Collins: New York).

[4] Lewis, C.S. (1947). *The abolition of man.* New York: Macmillan

[5] http://www.titanicstory.com/interest.htm.

[6] See: *http://www.youtube.com/watch?v=Fhb9bEEH76I.*

[7] Concerning differing company procedures, see: Lloyd's Calendar (1994). 'Company signals' and 'distress signals' of the steamship lines. *The Titanic Comutator, 18* (1) 45-48.

[8] see: http://www.titanic-titanic.com/warnings.shtml

[9] http://en.wikipedia.org/wiki/Blue_Riband

[10] see: http://www.titanic-titanic.com/titanic_myths.shtml. And see Maltin (2010) op cit.

[11] see: http://www.encyclopedia-titanica.org/

[12] Technically, this phenomenon is known in current human performance research as the "Soda Straw" problem. While providing detailed information with respect to one specific part of a display, this phenomenon trades a move general level of "situation awareness". And see: Smith, K., & Hancock, P.A. (1995). Situation awareness is adaptive, externally-directed consciousness. *Human Factors*, 37 (1), 137-148.

[13] see: Poulton, E.C., Hitchings, N.B., & Brooke, R.B. (1965). Effect of cold and rain upon the vigilance of lookouts. *Ergonomics, 8,* 163-168. See also: Hancock, P.A. (1984). Environmental stressors. In: J.S. Warm (Ed.). *Sustained Attention in Human Performance.* (pp. 103-142), New York: Wiley. See also: Hancock, P.A. (2013). In search of vigilance: The problem of iatrogenically created psychological   phenomena. *American Psychologist, 68* (2), 97-109. It is also feasible that the lookouts might have been subject to a limitation known as night myopia: see Leibowitz, H.W. (2002). The symbiosis between basic and applied research. In: J. Andre, D.A. Owens, and L.O. Harvey, Jr, (Eds.). *Visual perception: The influence of H.W. Leibowitz.* Americna Psychological Association, Washington, DC., as well as: Levene, J.R. (1965). Nevil Maskelyne FRS and the discovery of night myopia. *Royal Society of London Notes and Reports, 20,* 100-108.

[14] Those were indeed heroes who did their duty to the end and it is thus important to list their names. They were: Theodore Ronald Brailey (Pianist), Roger Marie Bricoux (Cellist), John Frederick Preston Clarke (Bassist), Wallace Hartley (Bandmaster), John Law Hume (Violinist), Georges Alexandre Krins (Violinst), Percy Cornelius Taylor (Cellist), and John Wesley Woodward (Cellist).

[15] The calculations are as follows: Titanic weighed a reported 52,310 tons and succumbed in approximately 2 hrs. 20 mins; being some 8,400 seconds for a rate of 6.3 tons per second. Hindenburg weighed 237 tons and succumbed in 37 seconds giving a ratio of 6.4 per second. Of course, these rates are necessarily approximations, and much argument can be had over Titanic's time to sink which is often noted as 2 hrs 40 minutes. Similar issues derive from Hindenburg since cameras did not capture the first moment of ignition. Also, we do not have precise gross tonnages for either vessel as the respective disasters occurred. Thus, the estimates are necessarily approximations and not precise numbers. Presumably, I could conjure exactly the same number, however, this is pushing coincidence beyond reality. In truth, the whole identification is a form of *cognitive apophenia*. It is perhaps also of interest that the *Titanic* weighed approximately the same as the fully laden Edmund Fitzgerald, and see: MacInnis, J. (1998). *Fitzgerald's storm: The wreck of the Edmund Fitzgerald*. Thunder Bay Press: Thunder Bay.

[16] There was at least one male passenger who survived the disaster by dressing as, and pretending to be a woman (Maltin, 2010 op cit).

[17] Frey, B.S., Savage, D.A., & Torgler, B. (2011). Who perished on the *Titanic*? The importance of social norms. *Rationality and Society*; 23(1), 35-49. And: Beavis, D. (2002). *Who sailed on Titanic? The definitive passenger list*. Hersham: Ian Allan Ltd.

[18] Of the *Hindenburg's* 36 passengers some 23 survived, giving a survival rate of 63.8%. On the *Titanic* were 329 first class passengers of whom 199 survived giving a survival rate of 60.5%. These values are quite impressively close together, however, the breakdown by age and gender belies this simple equivalency.

[19] see: Spufford, F. (1997). *I may be some time*. St. Martin Press: New York. See also: Cherry-Garrard, A. (1922). *The worst journey in the world*. Carroll & Graf: New York.

[20] Recently, the violin of the band leader, Wallace Hartley, was auctioned off for approximately $1.9m (*Mail on Sunday*, October 20, 2013). All eight members of the band perished in the disaster, although unlike others, their bodies were all recovered. Hartley's body was found by the CS Mackay-Bennett and taken to Canada. Eventually, he was interred in his home town of Colne in Lancashire, England. Traced through his fiancée Maria, who had given the violin as a gift, its recent re-discovery has featured this artifact as the epitome of courage and perhaps even the quintessential symbol of the disaster itself.

[21] See: Mariott, P.B. (1992). *RMS 'Titanic:' Reappraising of the evidence relating to the S.S. 'Californian.'* Marine Accident Investigation Board, HMSO: Southampton.

[22] While Titanic went down on April 14th, 1912, one of her lifeboats (Collapsible A) was only recovered some 29 days later on May 13th. It was found drifting by the *Oceanic* with three bodies still in it, very probably the same three that Fifth Officer Lowe had left there when attending to the living that night in April.

[23] And see: Hancock, P.A. (2009) *Richard III and the Murder in the Tower*. History Press: Stroud.

"... the tallest structure ever built in northern Europe and certainly the most ambitious Cathedral project of the High Gothic era." [1]

5. REACHING FOR GOD

5.1. Gothic Creations

The Cathedrals of *Chartres* and *Rheims* are among the greatest buildings in all of human civilization. Their respective stories have been largely ones of sustained success. They still stand today, actively resisting the ravages of time and providing a ready and continuing experience of awe [2]. What I was searching for to complete my trinity of *air, water* and *earth* was a cathedral that had suffered in the same catastrophic way as *Hindenburg* and *Titanic* had suffered. It appeared, initially that there might even be several candidates. For example, one possible case was the original Cathedral of *Salisbury* or *Old Sarum* in central southern England. This cathedral has, for a number of reasons, almost completely disappeared. Today, the only remnants you can see are the remains of its ghostly footprint enshrined in the grass [3] (see Figure 5-1). However, the central theme of my work is one of sudden disaster at the edge of technological achievement and the old Salisbury Cathedral had been abandoned largely as a result of a slow disaster from not having either a sufficient or convenient fresh water source. For the epitome of sudden disaster I eventually found the quintessential example. So, here I want to relate the story of *Chartres* and *Rheims'* damaged cousin; that is, – the stunted wonder of *Beauvais*.

© Springer International Publishing AG 2017
P. Hancock, *Transports of Delight*, DOI 10.1007/978-3-319-55248-4_5

Figure 5-1. *The footprint of the Cathedral of Old Sarum sits like a bleached skeleton inside the large earthworks of the prior eras. (Photograph by the Author)*

5.2. The Incomplete Wonder

The never completed Cathedral of *Beauvais* still stands today on the plains of northern France. It represents what is arguably the greatest in a succession of structures that sought to touch God by reaching up into the heavens. The appearance, in their magnificent splendor, of these cathedrals of northern Europe represents one of the most interesting epochs of technological progress as well as one of the more mysterious stories of sudden

Figure 5-2. *The stunted but wondrous Cathedral of Beauvais in northern France*

brisance of civilization and sophistication in the whole of our human story [4]. It is not as though organized human society was not capable of generating such vast architectural achievements, as the seven wonders of the ancient world attested. [5] However, it is still not fully understood how a continent in the beginning throes of recovery from the nominal 'dark ages' could suddenly create such wonders. Historians and scholars point to a confluence of many factors, (but when, in present day discourse do they not)? It is clear however, that many of the great master masons who built these wonders were actually exploring architectural hypotheses as much as they were creating physical buildings. Further, many of these finished cathedrals materially represent a series of overlaid efforts by different succeeding master masons across multiple generations of effort. In this sequence of Gothic

Cathedral development, as we shall see, perhaps the most persistent theme could well be entitled '*ever higher.*' Indeed this might be very the motto of *Beauvais* Cathedral itself.

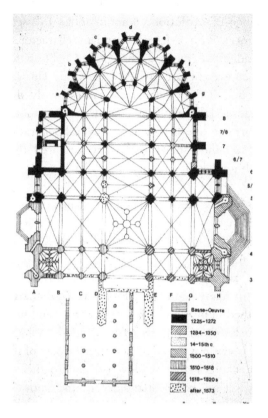

Figure 5-3. *The stunted magnificence of Beauvais Cathedral is seen in this schematic which shows the stages of completion over the centuries.*

Started around 1225, *Beauvais* was then the latest, and was designed to prove to be the greatest in the pantheon of such spiritual structures. It proved essentially to be the last in this sequence of great cathedrals that dominated the architecture of Western Europe in the centuries following the turn of the first millennium. *Beauvais'* mixed architecture is reflective of its place in this sequence of evolution. *Beauvais* looks to reconcile early Gothic influences with a later rayonnant character. The latter perspective sought to emphasize height and light as the architects looked to reach ever higher toward their God. This driving force was certainly one of the fundamental causes involved in *Beauvais'* eventual collapse. Like all such cathedrals, *Beauvais* then was a collective effort and one that spanned several generations. In this sense it is a sustained social vision and required the power of persistent faith to make it a reality.

The present day *Beauvais* Cathedral stands on an ancient site with suggestions of a founding in the third century AD by St. Lucien. Documentary evidence notes the presence of a Bishop Maurinus in 632 AD by which time some elaborated site of worship was already present in the immediate post-Roman settlement. Not many centuries later, *Beauvais* could boast a fully developed cathedral begun in the tenth century and finished in the eleventh. We know of this structure, called "*Notre Dame de la Basse-Oeuvre,*" because we can still see a major part of it today. The Basse-Oeuvre stands in the shadow of, and to the west of, the "*haute oeuvre.*" This latter structure is the more commonly recognized and slightly later cathedral, of which we see and shall hear much more. Sadly, during the construction of the world-famous building for which *Beauvais* is now rightly known, part of this earlier cathedral was demolished. Like its 13th century counterpart, the "*Basse-Oeuvre*" stands today only partly complete; a remnant of a now passed and greater glory. Over the centuries, this smaller wonder has itself come under repeated threat and its present existence, unique as it is, represents a small miracle of its own. However, it is one that is often neglected by visitors who can only 'see' the larger wonder.

The larger structure owes its existence in part to the destruction by fire of a portion of the former, smaller cathedral. Faced with the costs of repair and partial replacement, the decision was made to "*go for something bigger.*" The first overseer of new construction was Bishop Miles (Milon de Nanteuil) whose original conception was largely along strict Gothic lines [6]. His efforts were eventually stultified by lack of resources and problematic administrative issues (where have we heard that before)! In 1238, Bishop Robert assumed responsibility for the building's continuing development. He endeavored to cut his cloth according to his budget and changes were made in the design. However, despite this scaling down, *Beauvais* remained perhaps the most ambitious project of the age; determined to outshine all of its many peers. In the decades of the 1250's and the 1260's Bishop Robert's successor, Bishop William, succeeded. William added further height to make *Beauvais* the tallest of all European cathedrals. [7] Construction continued apace and had such dreams been fully realized today, we might well talk of *Beauvais* in the same hushed and venerating tones that we use for *Chartres and Rheims*. Tragically, at *Beauvais* the builders reached just too far. Their vaulting ambition proved to be a step too far into what eventually became a *Tower of Babel* in stone. Let us see what happened.

5.3. Vaulting Ambition

Before we proceed to an examination of the collapse itself, let us try to imagine for a moment, the effect of such a cathedral on the average member of local thirteenth century society. For many such individuals, even a two-story structure would be quite an unusual sight. Rooms inside buildings were almost generally small. They were also dark with low ceilings and narrow, dirty windows which were themselves still something of a luxury at that time. Colors were rather muted if not overtly drab. This was true in virtually every part of life, except for the wonders of nature itself, little wonder that faith, and often blind faith was such a driving force in their existence, especially when the muted capacity of humans was juxtaposed in the common mind to God's wonderful creation that surrounded them [8]. Loud noise, for example, a source of pollution in today's society [9], would be rare in the thirteenth century since people would not often collect in any large numbers. For the occupant then of the average hamlet a visit to any cathedral would have far outstripped the shudderingly exciting experience of even Disneyland for one of today's ardent seven-year olds. Just seeing the outside of a cathedral must have been the thrill of a lifetime; but inside [10]! Although no thundering organs were present in 1272 when services were first held, the reverberating sound of the choir may well have brought rough hands to ears. Not just the internal height, but the light and, from the stained glass windows, the blinding colors. Surely this must indeed be the House of God and perhaps even a very reflection of the celestial realm itself!

The sermons and morals presented inside and the proclamations outside made greater impacts and served as virulent dissemination of putative "information" more than any news channel could ever dream of achieving today. Little wonder that Victor Hugo could set his marvelous novel, featuring the *Hunchback* in and around *Notre Dame,* which still evokes a latent sense of awe so many centuries later. Weaving in the added dimension of 'sanctuary,' Hugo provides us with a passing glimpse of this world of wonder, a world truly beyond the secular, of which now we can only see the faintest echo. From the time they were built and for hundreds of years after, they remained the highest and most awesome structures anyone would encounter on the whole of the European continent. They were the pinnacle of twelfth and thirteenth century technology, reigning unchallenged for decades and even centuries. Their life as 'transports' continues unabated even in today's jaded world.

To illustrate this point visually, we can, for example, compare an early print of *Beauvais* (an example of which is shown in Figure 5-4) with a modern photograph (see Figure 5-5) taken from almost the same location. Such a comparison serves to show us that even centuries later, *Beauvais* still dominates and overshadows the mundane, everyday dwellings that surround it. It is hard to envisage that anything we build today will have such longevity or leave such an on-going legacy.

Figure 5-4. *Some conception of the power of Beauvais is provided in this print. Below are the everyday streets with their workaday characteristics. Above towers the splendor of the cathedral. Almost ethereal in its effect, the print shown captures this 'other' worldly nature of the cathedral at its time of building and even to the present. It is no wonder it is a 'transport of delight.'*

5.4. Another Night to Remember

In his most illuminating text, Murray provides us with an often quoted passage about the collapse of *Beauvais* [11]!

> *"On Friday November 29th, 1284 at eight o'clock in the evening, the great vaults of the choir fell and several exterior pillars were broken; the great windows were smashed; the holy chasses of Saint Just, Saint Germer, and Saint Evrost were spared; and the divine service ceased for forty years. Several pillars were interposed in the choir arcade in order to fortify it."*

Concerning theories as to why the collapse had happened, Murray himself is a little more discursive, but polemic, when he argues:

> *"Very few writers have been able to escape the deeply ingrained belief that there must be a link between the disaster and the status of the Beauvais choir as the tallest Gothic Cathedral ever constructed, and that the limits of the strength of materials and human ingenuity had been reached. In one respect, this belief is obviously incorrect; stone can be piled as high as a mountain, and yet the material at the base will not be crushed. It is naïve to consider the collapse as a punishment for the hubris of the builders. Yet there is an element of truth in the explanation linking sheer height to structural failure."*

Murray goes on to explain that the problem that was identified actually concerns building open spaces inside mountains. More recent architectural experimentation has suggested that a marginally greater height in stone might possibly have been attained. However, this would need a more modern design using curvilinear rather than rectilinear buttresses.

However, even the latter structure could not reach much higher than *Beauvais* now stands which very much approaches the theoretical maximum for those design constraints. In contrast to his first observation, Murray concluded:

> *"Thus, it is not inappropriate to make a connection between the fact that the Beauvais choir was the tallest cathedral ever constructed and that it collapsed."*

But height alone cannot be the whole reason for the crash. For as we have seen, part of *Beauvais* remains standing to this day, even if it is in an altered form and with greater numbers of supports. As with the nature of our necessary knowledge about both past and future [12], the problem is that we know frustratingly little about the collapse of 1284. There are a number of theories which focus, quite logically, on structural integrity. These theories each feature or emphasize one specific element, although the degree to which these various identified factors might interact still remains unspecified. One of the first theories focused on the issue of roof support. At the time of the collapse there were only about half of the number of roof supports that are now present. Still a popular theory today, this proposition has it that the architectural focus on increasing the overall area of the stained glass windows left the roof support structures dangerously and finally, fatally weak [13]. The stained glass in the cathedral is, of course, analogous to the 'display space' of the various forms of our modern media. However, when one could only provide 'static' picture representations, then obviously the larger and more startling the area displayed the better [14].

Figure 5-5: *The great massif of Beauvais towers above the surrounding habitat. How much more of a wonder this must have been when it was first constructed!*

It was in these stained glass windows that many of the elevating stories of the Bible could be "told" to a congregation, many of whom were illiterate. This theory pits the very arc of purpose of the cathedral against its basic structural integrity. It is emblematic of the whole present text that when the conditions in which the 'why' of purpose is out of balance with the 'how' of process, then disaster looms. Previously, I have proposed that *'purpose predicates process.'* That is, the why of any matter should dictate how the specified goal is achieved. However, reflectively *'process promotes purpose'* and thus how a goal is to be approached places a necessary constraint on whether that goal can actually be achieved, as well as the vision of possible future goals [15].

Beauvais illustrates the case when the symbiosis of these two human dimensions clearly tipped beyond balance.

Rather than problems and issues with the vertical supports, another competing theory points to the foundations themselves. In this account, the uneven settling of the structure is proposed to have place intolerable tortional stresses on the building. Such a proposition does carry weight, since several present day cathedrals also record such flaws in their own foundations. Their preservation and maintenance is accordingly very costly. But, again as we know, *Beauvais'* half-cathedral still stands today and a caveat to 'foundation' theory must be this longevity of the remaining structure, although of course it has been reinforced a number of times in the interim. Yet a third theory points to certain environmental influences, amongst which the de-stabilizing effect of gale force winds from across the English Channel have been blamed. I'm sure this is a popular theory in northern France where Anglo-Gallic confrontations have been played out over at least the last millennium. Foul winds from England must resonate well as an explanation in the ears of the French as the cause of their woes!

Yet a fourth theory points to the potential problem that came from the use of successive masons and architects, each trying to perfect his own individual vision, yet having to integrate his own ideas into an existing structure. In more than an echo of today, we might frame this failure as being a result of emphasizing on-time, on-budget completion demands interwoven with perhaps over-ambitious plans and insufficient funding. If this was truly the case, *Beauvais* is just as much a lesson for us today as it has been for over-reaching administrators throughout the centuries. Poor

Thibaud of Nanteuil, barely a year into his episcopate (1283-
1300), and spectacular disaster was visited upon his cathedral!
This was an age when such portents could well be instrumental in
people being burned at the stake! Construction on *Beauvais* ceased
for almost one-hundred and fifty years. There were naturally works
of maintenance but *Beauvais* had to wait until 1499 for the Chapter
to consider the time ripe to pursue the vision of a completed
cathedral [16]. The next natural extension to the then existing
structure was to complete the Transept and across the years 1500
to 1550 these impressive wings of the cathedral were completed
one after another.

It was in 1561 that the authorities took their next fateful
decision. Most probably for logistical reasons, it was decided to
complete the Tower and Spire over the center of the Transept, as
compared, for example, to the completion of the rest of the Nave.
A lantern tower, somewhat like that of Ely Cathedral in England,
was created and topped with a spire surmounted by a highly
conspicuous iron cross. Those additions made *Beauvais* the tallest
house of worship in all of Christendom, even outstripping *St.
Peter's* in Rome. Barely six years after its erection, the Iron Cross
was removed because of concerns about its weight and safety.
Beauvais hit its second "iceberg" on Ascension Day (April 30,
1573) when the tower itself collapsed. This collapse occurred just
after the congregation had left the building and no lives were lost.
But despite a half-hearted effort at further expansion in 1600, the
building phase of the cathedral was done [17]. Eventually, the
west end was permanently blocked off and this blank and highly
disappointing "wall" is what we now see today. Despite the
cathedral being hit by five bombs during World War II, the
wonder of Beauvais is still there for one to see. Piquantly, *Beauvais*

is as stubborn as she is beautiful. Its older forebear, the *"Bas Oeuvre,"* is still attached to the stunted *"Haute Oeuvre"* which only ever saw completion in the mind's eye.

When building a ship to challenge nature, some noted that it was unwise to use terms like 'unsinkable.' *Titanic* engineers themselves did not use the term. It was only the advertising department who flirted with such an epithet. Further, when some were heard to say *'God himself could not sink this ship,'* many of the religious persuasion saw the disaster in terms of the Lord's vengeful but justified anger. We have no comparable, contemporary commentary on the *Beauvais* disaster. However, it did not take the most erudite of clerics to link *Beauvais'* soaring vaults to the Biblical *"Tower of Babel."* The outcome therefore was also most probably seen as a just result for *"Man's challenge to God."* Indeed, perhaps this will always be the theological response to technical failure at the edge of innovation. It will ever be characterized as evidence of God's disapprobation of human reaching, seen as efforts to usurp heaven's prerogative. Like all *post hoc* interpretations, theological or otherwise, such observations are made to fit the prejudice of the observers themselves. These, very fortunately, always happen to coincide with God's own identified divine will [18]! I wonder what God actually thinks of all this?

5.5. How Hath the Mighty Fallen?

Before we complete the narrative concerning *Beauvais* it is important to understand that its story is not unique as it might at first glance appear. As the builders of these spiritual structures were always exploring the unknown and did not know how long their creations might stand, it is not unusual to find episodes of collapse or of failure. In respect of my favorite stories of dissolution, I am very attracted to the sad loneliness of *Abbey Dore*. Located in the wonderful 'Golden Valley' of the Welsh Marches in the English country of Herefordshire, *Abbey Dore* is very much like a small and forgotten *Beauvais* [19]. It is true that the main destructive cause at Abbey Dore was man and not God, yet the remaining structure is very similar to Beauvais. One has a Church with a Chancel and a Transept but no Nave. And even the still functional remains of *Abbey Dore* have now been almost abandoned in a progressively more secular world. There remain only a few stalwart champions to fight its cause. Much less neglected, yet no less incomplete is *Rosslyn Chapel* of "*Da Vinci Code*" fame. Here, we find another stunted wonder. Clearly developed for a specific sponsor, Rosslyn contains many magical features – in every sense of the word – yet it remains much less than was initially envisaged [20]. Our landscape is literally littered with these wreckages of past technology, as our own contemporary 'swap meets', 'garage sales', and 'flea markets' continue to show us.

5.6. One Magnificent Mess

I am going to conclude my story of *Beauvais* Cathedral by narrating my own personal experience of it. Indeed, I am actually writing this as I sit inside the structure itself. So at this moment, I am in *Beauvais'* choir. From the outside, the whole edifice looked like nothing so much as a modern-day building site. In truth, it is a thousand-year-old building site. It is a construction that has never been completed. Inside the cathedral it is exactly the same scene, being one of quiet and rather desperate of remedial efforts. I feel cheated, short-changed. What should be a whole wonder remains only a half-complete masterpiece, begging to be finished. It is a longing that has now gone unsatisfied for almost a millennium. At first, as you stare up at the height and think, this is just like other Gothic Cathedrals. After all, how could just a few feet in height make a difference? But, this is a difference that makes a difference, and what a difference it makes! Here, they reached higher toward God than anyone in all the generations before; and higher toward spiritual fulfilment than almost anyone since. This wreck is the stark emblem of what perhaps might be envisaged as almost inevitable failure. It is an ambivalent epitaph to such dedicated and inspired individuals who imagined its completion.

Failure tinges all of the *Beauvais* experience; it haunts you. Everywhere you see the evidence. Today, wooden beams criss-cross the transept, scaffolding obscures the Nave and iron bars tie together various buttresses, in an anachronistic marriage of ancient and modern. All those efforts search to knit and sustain the creaking fabric of this fading wonder together. Even the earliest structures are not immune. Pillars were added very early to shore up the initial design, and while it is true that they are added in an unobtrusive and even masterfully hidden way, they are evident for those who know where to look and they are needed. The outer isles of the apse are higher than the main

Figure 5-6: *The flawed wonder of Beauvais. (Photograph by the Author).*

Figure 5-7: *Even as one contemplates the sublime, the necessitated scaffolding scars one's perspective.(Photograph by the Author).*

achievements that exist in many medieval abbeys. *Beauvais* presents inspiration and dare I say it, vaulting ambition!

I stand, I wonder, I marvel. Could such a creation alone bring someone to a faith in a higher being? Certainly, the upward angled inclination of one's head, drawn up by the stupendous height, places you in the right attitude to face some celestial God. The remaining nave very much serves to make you aware of just how small and insignificant you are. Then you turn to look at the crippled, gable end and all at once the illusion is broken. Here, perceived height is not nearly so breath-taking. In fact, it proves to be almost mundane. The utilitarian requirement to block up the incomplete cross of the transept is anodyne to any transcendental temptation [21]. And one slowly then begins to see each individual wart on this vision of beauty. The temporality and the terrestriality of the structure are emphasized by the dust-covered French artisans engaged in their leisurely discussion of their prospective labors. The ultimate impression one receives is that of an ineffable sadness. It is emblematic of the necessary triumph of humility over hubris; of the material over the spirit, of physics over philosophy. It is Keats "*Ozymandias*" made painfully manifest in stone. It is the essential epitome of frail humanity. And it hurts.

Now, outside the cathedral, this quiet desperation becomes even clearer. This is most obviously attested to by the evermore intricate and interwoven buttresses, mandated by necessity and fabricated by despairing genius. The veritable sequoias of stone stand as mute testaments to the sustenance of the unsustainable. All around life goes on; the French kiss each other in the café across the street and appear to be virtually indifferent to one of the wonders of the world which calibrates their own transient existence. Prosaic but praiseworthy restoration continues. The upper reaches of the south transept effulge yellow-white in the afternoon sun. It is almost the bitterest of all possible reminders of just what this building might once have been. In the end the mute stones speak to me and I am forced, but yet happy, to listen. They speak of the material manifestation of the power of longing and, at last, of hope. Although this one particular hope stands sadly unfulfilled, I am forced to acknowledge the aspiration that underpins its essence and that of its peers. And now the bell tolls to draw me away. I have been but a fleeting shadow in the river of time that has carried *Beauvais* down the ages [22]. I will take its sadness with me, I will distill its lessons and I will look to provide it just the smallest iota of comfort and assuagement – for me it has fulfilled its role as my final, wonderful, if flawed "*Transport of Delight.*"

Reference Notes: Reaching for God

[1] Murray, S. (1989). *Beauvais Cathedral: Architecture of Transcendence.* Princeton University Press: Princeton, NJ.

[2] see for example: Swaan, W. (1984). *The gothic cathedral.* Park Lane: New York; and: Calkins, R.G. (1998). *Medieval architecture in western Europe from A.D. 300 to 1500.* New York: Oxford University Press.

[3] Like the original Cathedral at *Arras,* that of *Old Sarum* has now almost completely disappeared; and see: Old Sarum Cathedral (1846). *The Ecclesiologist,* Vol. 3, 60-62, Cambridge: Cambridge Camden Society.

[4] Aubert, M. (1959). *Gothic cathedrals of France and their treasures.* London: N. Kaye.

[5] The seven wonders of the ancient world were respectively: (i) The Great Pyramid of Giza, (ii) Hanging Gardens of Babylon, (iii) Temple of Artemis [Diana] at Ephesus, (iv) Statue of Zeus at Olympia, (v) The Mausoleum of Halicarnassus, (vi) Colossus of Rhodes, (vii) Lighthouse of Alexandria. And see also: Ash, R. (2000). *Great wonders of the world.* Dorling: Kindersley.

[6] see: Forstal, J., & Magnien, A. (2005) *Beauvais Cathedral.* Picardy Department of Culture: Americas.

[7] Beauvais of course, does not reach the greatest single point height which tend to co vary with the height of any particular spire. For example, Salisbury (the new Cathedral) reaches a height of 404 ft. The "historical" in Beauvais refers to the height of the choir.

[8] And see the arguments in: Sheridan, T.B. (2014). What is God?: Can religion be modeled. New Academia Publishing, Washington, DC., and Moray, N.P. (2014). *Science, cells, and souls,* in press.

[9] Szalma, J.L., & Hancock, P.A. (2011). Noise and human performance: A meta-analytic synthesis. *Psychological Bulletin, 137*(4), 682-707. And see also: Szalma, J.L., & Hancock, P.A. (2012). What's all the noise? Differentiating dimensions of acoustic stress and the limits to meta-analysis. *Psychological Bulletin, 138*(6), 1269-1273.

[10] Such wonder and awe concerning physical structures can well present individuals with what the late philosopher, Paul Kurtz called the transcendental temptation. See: Kurtz, P. (1991). *The transcendental temptation: A critique of religion and the paranormal.* Prometheus Books. Amherst: New York.

[11] Murray, S. (1989). *Beauvais cathedral: Architecture of transcendence.* Princeton: Princeton University Press, (pp. 112, 115).

[12] Hancock, P.A. (2013).*On the symmetricality of formal knowability in space-time.* Paper presented at the 15th Triennial Conference of the International Society for the Study of Time, Orthodox Academy of Crete, Kolymbari, Crete, July.

[13] And see: Heyman. J. (1967). Beauvais Cathedral. *Transactions of the Newcomen Society*, 40, 15-32. And also: Wolfe, M.I., & Mark, R. (1976).The collapse of the vaults of Beauvais Cathedral in 1284. *Speculum, 51* (3), 462-476.

[14] see, e.g., Nass, C. & Moon, Y. (2000). Machines and mindlessness: Social responses to computers. *Journal of Social Issues, 56* (1), 81-103.

[15] Hancock, P.A. (2009). *Mind, machine and morality.* Ashgate: Chichester.

[16]   See: http://www.360cities.net/image/beauvais-cathedral-ambulatory#427.53,-0.00,15.0.

[17] http://www.frenchmoments.eu/beauvais-cathedral/

[18] Hancock, P.A. (2014). *Hoax springs eternal: The psychology of cognitive deception.* Cambridge: Cambridge University Press.

[19] Shoesmith, R., & Richardson, R. (1997). (Eds.). *A definitive history of Dore Abbey.* Herefordshire: Logaston Press.

[20] Butler, A., & Ritchie, J. (2006). *Rosslyn revealed: A library in stone.* O Books: London. And for the mythology surrounding that structure, see: Wallace-Murphy, T., & Hopkins, M. (2000). *Rosslyn: Guardian of the secrets of the Holy Grail.* Barnes & Noble: New York.

[21] Kurtz, P. (1991). *Op cit.*

[22] Thomas Campbell. (1777-1844). *The river of life.*
http://www.poemhunter.com/poem/river-of-life-the/
http://en.wikipedia.org/wiki/Thomas_Campbell_%28poet%29

❖❖❖

"Cathedrals, Luxury liners laden with souls,
Holding to the east their hulls of stone." [1]

6. SURVIVING SISTERS

6.1. Introduction

Up to this point I have focused upon the specific stories of the
three 'transports' which each suffered a sudden and disastrous
destruction. One obvious reason for this is that the stories
associated with the *Hindenburg*, the *Titanic* and to a lesser extent
Beauvais Cathedral, are well-known. As accounts of disasters, they
almost naturally attract our interest and attention [2]. However,
for the purpose of my overall theme, it is almost equally
important to examine and feature the stories of the peers of these
respective 'transports,' which either did not fail in so catastrophic
manner, or experienced their eventual dissolution in some much
less spectacular fashion. So, in this chapter, I consider the stories
of these "surviving sisters." To do this, I have examined a number
of companion structures. As we shall find, one or two of these
have interestingly, suffered in similar fashion to their more famous
siblings.

6.2. Chartres Cathedral: Beauvais' Living Cousin

In looking for *Beauvais'* "surviving sisters" I was able to choose
from among any number of French and English Gothic
Cathedrals, each of which possess several centuries of narrative of

© Springer International Publishing AG 2017
P. Hancock, *Transports of Delight*, DOI 10.1007/978-3-319-55248-4_6

Figure 6-1. *The full magnificence that is Chartres.*

their own [3]. However, having read my introduction, it will come as no surprise that I have chosen my favorite: *Chartres*. To try to recount the full story and history of *Chartres* in one brief chapter is essentially an insuperable challenge. The stories of *Titanic* and *Hindenburg* have been played out in many recent films, movies, and TV specials. The result is that many know these latter stories and, as these are relatively 'recent' events, we still have many artifacts, records, eye-witness accounts, and even visual recordings to inspect [4]. The origin of *Chartres* lies so deep in history that we have little in the way of reliable information about its earliest years. Of *Chartres*, Julius Caesar reported that on the same day each year, Druids used to sit together in this sacred place to address the problems brought to them by people from all over the local countryside. It is our earliest formal reference to *Chartres*.

There exists some evidence suggesting the presence of an altar on the site of *Chartres*, dedicated to the Virgin Mary, as far back as the sixth century AD. A royal decree in the eighth century AD also refers to a *St. Mary of Chartres*. However, the first unequivocal evidence we possess of a 'cathedral' appears around 743 AD when records report that it was actually 'destroyed.' Such destruction happened during a regional struggle when the structure was set on fire by the Duke of Aquitaine [5]. While the *Chartres* we see today appears so ancient that it is tempting to conclude that it has always been as such, in fact this house of worship has been built, destroyed and reconstructed on at least seven occasions. Hence, it is the *perceived* sacred nature of the site and the *idea* of a cathedral that have actually proven more persistent and durable than any physical representation. It is this ideational permanence which means *Chartres* still stands today in an age where trans-Atlantic airships and trans-oceanic liners now seem almost ultimately anachronistic. However, as we shall see, the best way to express this thought is that it is the idea itself that persists. But now trans-oceanic travel is sub-served by "newer" technologies. I shall eventually question the conception of "newness" and what exactly *"making it new"* actually means.

Chartres was destroyed yet again in 858AD by a Viking invader named 'Hastings' [6] but, like the phoenix it is, it rose again when Charles-le-Chauvre sponsored a new church to be built under the direction of Bishop Gislebert. Charles gave the Virgin's veil to this resilient house of God and it remains one of the most revered relics in the cathedral's possession, and indeed in all of Christendom. Little wonder that *Chartres* was considered the European center for the worship of Mary, the Mother of God. This particular church lasted until 1020 AD when again it suffered from the ravages of fire. It was at this time that St. Fulbert (960-1028 AD) asked his architect Beranger to help rebuild while Fulbert himself helped enlarge the crypt. Just over a century later, in 1134 AD, the town was again set on fire and Fulbert's cathedral facade was damaged. Rebounding once more from this setback, more building efforts were enacted and the world famous 'Royal Portal' was begun in 1150 AD. Work had been progressing for some four decades when again the whole construction was damaged by fire on the night of June 10th, 1194 AD [7]. One senses a theme developing here! Clearly, disaster does not restrict itself when it comes to cathedrals. Like many incidents of disaster however, *Chartres* clearly brought out supporters in their droves to help with the rebuilding. The parts of the church spared by the fire were incorporated into the emerging structure and this is much of what we see today.

Despite this ebb and flow of fortune, *Chartres* Cathedral remains one of the oldest and most revered shrines in all of Christian Europe. It was erected over earlier sacred sites. The apse, the nave and the transept we see started in 1194 AD and today stand over the more ancient crypt. The 'king's door' and the west tower also date from the 12th century. The cathedral was finally consecrated in 1260 AD. The years between 1200 AD and 1240 AD had seen the completion of the stained-glass windows which remain one of the great wonders of the world today. As with all Gothic Cathedrals, their design, architecture and construction were overseen by the guild of master 'masons.' However, completion was not possible without the collective effort and belief of both Christian volunteers and the local populace. In fact, these times seem to have been particularly special in that this building phase even required the suppression of local disputes and quarrels for the work to go forward. As Robert de Torigny, Prior of Mont Saint-Michel noted *"For the first time in Chartres one could see some of the faithful harnessed to cart-loads of stones and wood, and pulling carts laden with grain and whatever else was required for the building of the cathedral, the towers of which soared up as if by magic. Never would such wonders be seen again."* But of course, in their various forms, such 'transports' are evident in every generation.

If accounts of the *Hindenburg*'s failure can be framed in seconds and that of the *Titanic* essentially in minutes *Chartres'* success is one that has spanned centuries and can even legitimately now begin to be measured in millennia. *Chartres* is a 'thousand year technology.' In this, *Chartres* and all its companion cathedrals are unlike my other transports of delight in that they have outlasted all of their inaugural passengers and today, they continue their voyage and take on new passengers, each of a different generation. But this difference is the mere illusion of Protagoran hubris [8].

6.3. Graf Zeppelin: The Successful Airship

While *Beauvais* provided me with several possible 'sisters,' *Hindenburg* has only one natural sibling. That aircraft was the *Graft Zeppelin* and this was the 'transport' I naturally chose. Even while recognizing that *Hindenburg* had previously had a successful voyaging season in 1936 it can, of course, become easy to focus on the dimension of risk, especially given our hindsight-based knowledge of the final disaster itself. In fact, the iconic motion pictures and still images surrounding the explosion remain our quintessential representation of lighter than air transport. Arguably, it proved to be these images that served to end the whole genre of such technical progress altogether. This may have been an underserved fate for such a technology.

However, hidden in the shadow of these famous flames was actually a very successful counterpart, the quiet elder sister of *Hindenburg, Graf Zeppelin* [9]. Some facts serve to reinforce the notion that long-distance travel by air-ship had actually been really quite safe. In nearly nine years of uninterrupted service, the 776 ft. long, *Graf Zeppelin* flew some one million largely trouble-free miles composed of some 590 separate flights. In those years of service, it carried more than 34,000 passengers with no appreciable injury of any sort. *Graf Zeppelin* regularly crossed the Atlantic; and actually made over 130 crossings in total. Powered by five, 550 horsepower Maybach engines, *Graf Zeppelin* was capable of carrying 15,000 kg at a speed of 80 mph and was the first aircraft to make a commercial round-the-world voyage. Officially launched on July 8th 1928 it was christened by the daughter of Ferdinand Graf von Zeppelin himself. The airship began her flight trials under Hugo Eckner who, as we have seen, would be at the heart of the German investigation of the *Hindenburg* crash. *Graf Zeppelin's* final test before entering full service was an extended thirty hour plus flight. This was shortly followed on October 11th 1928 by her first commercial flight which crossed the Atlantic. Departing Friedrichshafen, she arrived at Lakehurst some four and a half days later. As sometimes proved to be the case, the crew on this first trans-oceanic voyage outnumbered the passengers. Among the latter was U.S. Naval Officer Charles Rosendahl, an interested military observer who would later feature in *Hindenburg's* demise. Also aboard the *Graf Zeppelin's* inaugural commercial journey was Lady Grace Drummond-Hay acting as a reporter of the event to a number of news services. *Graf Zeppelin* had suffered some damage on that trip, but again this was at that time still in the nature of an experimental technology. The return flight, now with the jet stream helping, took just under three days. The first of many successes was now "in the bag" as it were.

Figure 6.2: Graf Zeppelin

Less than a year later, the ground-breaking and unprecedented round-the-world flight was begun on August 7th, 1929. More than 25,000 miles would be covered in just over three weeks. On this flight, *Graf Zeppelin* called at Tokyo and Los Angeles in addition to its standard "ports" of Lakehurst and Friedrichshafen [10]. *Graf Zeppelin* went on to set an even more startling series of precedents. In 1931, she made a three day polar flight carrying research scientists from multiple countries and more than 50,000 'letters' for stamp collectors eager to add this unique event to their collections. The summer of 1931 saw a regularly scheduled trans-Atlantic service between Germany and South America and in the following year this service encompassed eighteen successful round trips. In 1933, the theme of specialty trips continued with a visit to the Chicago World's Fair, by which time politics had begun to exert its inevitable influence. *Graf Zeppelin* carried the traditional three-colored German flag on one side of the fin and the Swastika on the other. It was a foreshadowing of *Hindenburg's* later appearance. Most interestingly, the crew apparently chose to fly in a clockwise direction round the Fair apparently in order to keep the Swastika emblem partially 'hidden.'

Following the rise of the Nazi party to power, in Germany ever greater use was made of the airship for such overtly political purposes. Hitler ordered that the *Graf Zeppelin* fly over Berlin as part of the May 1st, 1933 'May Day' celebrations. Later that month it flew propaganda minster Joseph Goebbels to an early Axis meeting. He gave the Italian Air Minister, Italo Balbo an aerial tour of his own capital Rome, in order to emphasize the superiority of German science and technology. A few months later the airship was again used as a technological appendage to one of Hitler's appearances at Nuremberg [11]. The Nazi Party was clearly very aware of the power of wonder involved with the appearance of such a magical "transport." This exploitation demonstrates to us, in rather clear and stark terms, that 'transportation' does not always mean that we end up in a better place. [12]

The year 1934 saw the airship now back to its main purpose of passenger travel and there was regular service to South America every fortnight. This pattern persisted in the carrying of both passengers and mail to Brazil on a regular and reliable basis. Thus, *Graf Zeppelin* had been in faithful and near flawless service for close to a decade, when in May 1937 she received news of the *Hindenburg* disaster while in mid-flight. On May 8th, she landed at Friedrichshafen and from that time never carried a paying passenger again. Her final flight was on June 18th 1937 when she was trans-shipped to Frankfurt and with all hydrogen removed was put on exhibit. In March 1940, Herman Goering's Luftwaffe dismantled her. *Graf Zeppelin* thus proved every bit as much a victim of the *Hindenburg* disaster as *LZ 129* herself. Whatever her previous record, *Graf Zeppelin* was now viewed through the same lens of history as *Hindenburg*. Sadly, that lens had fire and destruction imprinted all over it. *Hindenburg's* successful sister suffered a sad and ignominious end, and a fate that was very much undeserved for such a wonderful and reliable creation.

6.4. *Titanic's* **Tragic Sisters:** *Olympic* **and** *Britannic*

Although by far the most well-known, *Titanic* was only one of three sister ships and her sibling, the RMS *Olympic* had actually been the first of the three to begin construction. *Olympic's* keel was laid at the Belfast shipyards of *Harland and Wolff* on December 16th 1908 [13]. She was launched less than two years later in October 1910 and was officially completed on May 31st 1911 for a total cost of some 7.5 million pounds. Piquantly, or perhaps even purposively, this official finishing date was the same day that *RMS Titanic* herself was launched on to water. Some two weeks later in mid-June *Olympic* embarked upon her maiden voyage. Like *Titanic*, *Olympic* experienced early troubles when, on September 20th she collided with the Royal Naval Cruiser, *HMS Hawke*. The impact left gashes both above and below *Olympic's* waterline and her subsequent repairs affected *Titanic* in, perhaps critically, delaying the latter's maiden voyage.

Figure 6.3: *Shown here is the damage to*
Olympic after the collision with the HMS Hawke.

Following *Olympic's* repairs she returned to service on November 30th 1911. However, that was not the end of her troubles since she than subsequently lost a propeller to a submerged object. In an on-going litany of ill-luck, *Olympic* hit bottom and had to be inspected by divers for damage, although on this occasion no appreciable problems were found. As with the *Graf Zeppelin*, *Olympic* was in service when her sister ship went

down on April 14th 1912, being herself on a run from New York
to Southampton. Ignoring the callous call for her to trans-ship the
now shocked and battered passengers from *Carpathia* in order to
transfer them to New York, *Olympic* was quietly retro-fitted with
a series of additional collapsible life-boats, although by then, in
the eyes of the public, the damage had already been well and truly
done. Unrest was evident in the crew since boiler-room personnel
subsequently went on strike, feeling that even these steps toward
safety were simply inadequate to ensure the safety of the lowly
crew members.

Trouble seemed to ever dog *Olympic*. In early July, she lost
steering control and ran aground but was able to free herself with
the inevitable concomitant necessity for repairs. In October a
more thorough refit addressed the issue of the irregular height of
water tight compartments and safety was further improved with
the addition of more lifeboats and a double skin. It was only in
April the following year of 1913 that she returned to full service.
She continued to experience the rough and tumble of operations
when a wave shattered glass on board injuring some passengers. In
October, *Olympic* helped rescue the crew of a ship damaged by a
mine but the war of 1914 saw her laid up due to the inherent
danger of being such a prime target. However, just less than a
year later, she was requisitioned as a troopship and saw both
service and action in this capacity. In May 1918, she turned
heroic, ramming and sinking U-103 and continued in war service
until July 1919 [14]. By late June of 1920, she was back on the
New York run in her original incarnation. From a conduit of naval
power, she went back to her primary function of civilian transit.
Successful years followed until 1924 when she collided with the
Fort St. George and again had to undergo extensive repairs. It was

following the merger between White Star and Cunard when she culpably hit the *Nantucket Lightship* killing seven of the latter's eleven crew members [15]. Her last paying voyage was in March 1935 and in September of that year she was, like the *Temeraire*, sold to ship-breakers in the impoverished Jarrow area of Northern England. Her destruction was thus partly designed to provide jobs for an area deep in the throes of recession. By the end of that year *Titanic's* older sister was also gone forever.

Nearly two years after *Titanic* had slipped beneath the waves her other sister, *Brittanic* was launched. Almost immediately, the necessities of war mandated her requisitioning as a hospital ship with more than 3,000 beds. Her mission changed from one of providing delight to one of providing succor. Both these functions however, are sub-served in the notion of providing hope. Although it can be said that medical respite and treatment can certainly be a journey back from pain to pleasure, the origin in damage and injury has more a physical than spiritual wellspring. Under the auspices of the Red Cross, she made her maiden voyage to Mudros two days before Christmas, 1915. Following the war she was released back to the White Star Line and was in the midst of re-fitting back to passenger service when the government changed its mind and put her back into hospital service again. In November she experienced an emergency by running into a sea-going mine-field. During the evacuation thirty passengers were killed by the still rotating propellers.

During this event, *Brittanic's* starboard bridge was damaged and in a related failure her water-tight doors failed to function properly. Her captain, cognizant of the unrecoverable nature of a deep-water sinking, looked to intentionally beach the ship [16].

However, in this endeavor he was unsuccessful. One hour after the initial explosion, the ship keeled over on her starboard side and she sank in the Kea Channel [17]. One piquant codicil to her sinking was the story of one Violet Jessop, a young lady who had the dubious honor of going down with both *Titanic* and *Britannic*. In a coincidence testing all credulity, she was also on *Olympic* when it collided with HMS *Hawke*. However, it is comforting to report that Violet survived all three incidents [18].

6.5. Summary and Conclusion

The stories of the surviving sisters such as *Chartres*, *Olympic* and *Graf Zeppelin* tell us much about the initial failures of their more famous siblings that I previously recounted. Indeed to a degree, they provide the more normative backdrop to the unusual and even unique incidents of failure of their more famous relations. As we have seen however, not all of these latter stories were 'sweetness and light.' As hazardous technologies each on their respective 'cutting edge' of advancements in their own areas, it is actually rather unsurprising that we see such issues and problems. As the incarnation of the 'thousand year technology,' it is rather obvious that Gothic Cathedrals, in all in their various incarnations, will face challenges from fire, weather, and the depredations of time across the centuries. In epidemiology we might refer to this as a "mere exposure" effect to time itself. What the sisters' stories tell us is that each of these 'transports' clearly worked. They each achieved their primary function and the manifest goals intended. That they also served to transform lives and to transform societies beyond these intentional goals attests to the additional influences that technology always exerts upon all of us.

Reference Notes: Surviving Sisters

[1] The quote here is from Wystan Hugh (W.H) Auden.

[2] See; Petroski, H. (1992). *To engineer is human: The role of failure in successful design.* Vintage: New York, and also: Perrow, C. (1999). *Normal accidents: Living with high-risk technologies.* Princeton University Press: Princeton, NJ.

[3] Wilson, C. (2005). The gothic cathedral: The architecture of the great church 1130-1530. Thames& Hudson: London.

[4] see: http://www.youtube.com/watch?v=F54rqDh2mWA.

[5] Ball, P. (2008). *Universe of stone: Chartres Cathedral and the invention of the gothic: A biography of Chartres Cathedral.* Bodley Head: New York.

[6] http://aladyinfrance.com/history-chartres/

[7] Ball, P. (2008), op cit.

[8] Hancock, P.A. (2010). The battle for time in the brain. In: J.A. Parker, P.A. Harris, and C. Steineck (Eds.). *Time, Limits and Constraints: The Study of Time XIII.* (pp. 65-87), Brill: Leiden.

[9] http://www.airshipsonline.com

[10] http://www.airships.net/blog/graf-zeppelin-round-the-world-flight-august-1929.

[11] http://www.airships.net/lz127-graf-zeppelin/history.

[12] As we shall see in the next chapters, the very word "transportation" carries certain connotations with it, especially to the English. As a way to deal with progressive over-crowding, a strategy was enacted to force emigration by using minor crimes (and even some major ones) as an excuse. The United States provided the first "transportation" destination that after the American Revolution that avenue was closed off and England began to send people to Australia. The exodus continued even after World War II when such voluntary emigrants were known as "five pound pons," after the cost of their passage. For some, despite their being considered a "punishment," the opportunity which transportation offered proved to be of great benefit. See Hughes, R. (1987). *The fatal shore*. New York, N.Y.: Knopf.

[13] http://www.greatships.net/olympic.html.

[14] McCartney, I., &; Mallmann-Showell, J.(2002). *Lost patrols: Submarine wrecks of the English Channel*. Penzance: Periscope Publishing Ltd.

[15] http://en.wikipedia.org/wiki/United_States_lightship_LV-117.

[16] Powell, R.K. (1991). The final journey of the Britannic. *The Titanic Commutator, 15* (3), 19-25.

[17] Chirnside, M. (2004). *The Olympic-class ships*. Stroud: Tempus.

[18] Maxton-Graham, J. (1997). *Titanic survivor*. Sheridan House: Dobbs Ferry, New York. It is of more than passing interest that, statistically, there is likely to be at least one such person (as with the general surprise about the coincidence of birthdays in even relatively small meetings). A recent set of tragedies confirm this propensity. For example, the air stewardess Tan Bee Joek was supposedly going to be on both Malaysian Airlines (MH370) which was lost and MH17 which was shot down over Ukraine. One might say this is not so surprising since she was a crew member of the common carrier. However, there was also a passenger, Maarten de Jonge, a professional cyclist who was booked on both flights but eventually caught neither.

❖❖❖

"A labyrinth is a symbolic journey . . . but it is a map we can really walk on, blurring the difference between map and world." [1]

7. THE RIDDLE OF THE LABYRINTH

7.1. Augmentation to Locomotion

I realize that in what I have written so far, there is a danger of seeing the 'transports' I've described as essentially passive experiences. That is, they appear to be journeys that happen to you, or are imposed upon you, rather than you generating the experience yourself. Nothing could be further from the truth. Moments of wonder are always 'embodied' experiences [2]. That is, they necessarily require your active presence, and your volition, and your physical participation. As we have seen, the most apparent distinction between Gothic Cathedrals and trans-Atlantic ships is that the latter move you from origin to destination, from one physical location to another. In technical terms, ships, and airships are orthotic enhancements to locomotion. Like cars, wheelchairs, skis, sleds, wagons, bicycles, jet-skis and all their movement augmenting peers, they each add to the inherent human capability to crawl, toddle, walk, run and sprint from A to B. At first glance it appears that, apart from the procession itself, one very much ends up where one chooses to "dock" within any Cathedral. In short, pews seemed fixed and your seat appears to be a very static location. However, because of the riddle of the labyrinth this is not always necessarily so.

© Springer International Publishing AG 2017   113
P. Hancock, *Transports of Delight*, DOI 10.1007/978-3-319-55248-4_7

7.2. The Pagan in the Sacred

Embedded in the floor of the naves, the entries, and occasionally
the transepts of many cathedrals are (or were) ground labyrinths
or mazes [3]. These took many differing forms as we can see from
what is illustrated in Figure 7-1. Composed of a series of
concentric circles or other geometric designs, the fundamental
underlying image is founded upon the Cross and recapitulates
Christ's Crucifixion. It has been further suggested that these
constructions served to represent the signature glyph of each
master mason who was responsible for the respective cathedral's
construction. While this suggestion is interesting and may hold
some degree of credence, this was neither the sole nor the main
purpose of these fascinating creations. One symbolic
interpretation of such labyrinths are that they represent the "web'
of our transient, temporal world which, although spiritually and
morphologically founded on the Cross of Christ, actually serve to
represent the series of twists and turns, involved in the transient
course of human existence. Like life itself, each labyrinth has an
entry point which is representative of birth and a termination
point symbolic of death. Reminiscent of the Greek myth of
Ariadne, we cannot find our way through this maze of life alone
without external assistance, which in the present case of Gothic
Cathedrals is the divine Grace of God.

Figure 7-1: *From left to right and top to bottom, the images of the i) Chartres Labyrinth, ii) the St. Omer Labyrinth, iii) the Amiens Labyrinth, an; iv) the Rheims Labyrinth. (Drawings by Richard Kessler, Copyright © the Author).*

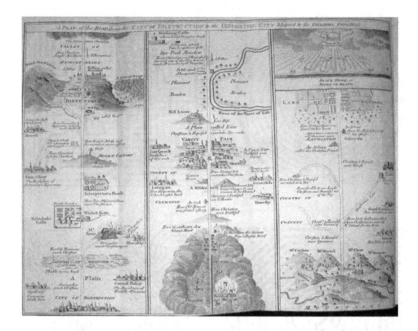

Figure 7-2: *The map of Pilgrim's progress laid out as an Ogilvy linear map.*

Every single one of these journeys has its own unique aspects as each individual peregrination proceeds. However, like Dante's allegorical rescue from Hell and Purgatory, with the right guide we can reach the goal of our journey [4]. The self-same notion of progress through a series of trials, distractions and traps is intrinsic to the *"Pilgrim's Progress"* of John Bunyan [5]. Indeed, the map of Pilgrim's progress looks very much like a labyrinth laid out in a linear fashion, as is shown in Figure 7-2 [6].

Figure 7-3: *Individuals engage in paradoxical peregrinations in Charters' Labyrinth.*

The forces of evil here are represented by the various distractions along the route. Traps like *'vanity fair'* and the *'slough of despond'* serve to de-rail those who then fall short of the fulfillment of spiritual life here on Earth and then beyond.

In the case of *Chartres*, perhaps the icon of evil is represented, at the center of the labyrinth by the Minotaur. This representation of the "beast" derives from the same Greek myth of Ariadne, whose story similarly features the actions of the hero, Theseus. In Dante's work, his eventual guide proves to be Beatrice who leads Dante up through the celestial circles of Heaven. It is surely no coincidence that the now lost central 'plate' of the *Chartres* Labyrinth then depicted the mythical battle between Theseus and the Minotaur [7]. Either directly, or by various forms of indirect implication, each of these labyrinthine representations illustrate how the spiritual path serves to resolve the surface chaos of the world with the purported perfection of the spiritual journey. But as I stated in the beginning of this chapter, these Labyrinths were not passive pictures just for unmoving contemplation. Rather, as we can see them from Figure 7-3, labyrinths were for walking. So, the Labyrinth presents an actual physical journey that has to be undertaken.

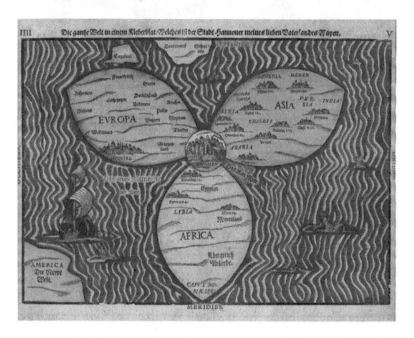

Figure 7-4: *Bunting's World Map showing Jerusalem as the epicenter of the world.*

7.3. Petite Pilgrimage

The paths of each labyrinth lead to their respective centers which in their turn, represent both the physical and spiritual *omphalos* of the Universe. It has been observed that almost all formal religions contain some teaching that emphasizes the metaphor of a journey [8]. In the Christian world, the terrestrial goal for this worldly pilgrimage was Jerusalem. We can see the emphasis of the centrality of Jerusalem in other documentation. This is true for especially early maps in which Jerusalem was shown as the focal point, see Figure 7-4. [9]

Either as a dimension of a formal indulgence or as an acceptable substitute for those who were lame, infirmed, or simply too poor for the real thing; walking the Labyrinth could be considered as a substitute or even an equivalence to an actual pilgrimage to the Holy City itself. In search of either a blessing or the expiation of sins, somewhat like the more arduous real-world journey, the labyrinthine surrogate could serve to cleanse and elevate the pilgrim's secular body in the same way that as it supported their spirit's search to embrace God's perfection. Such pilgrimages, in all their varying forms, foreshadow my later argument that any physical journey (representative or otherwise) is but a pale shadow of its internal equivalent. For, in this latter personal odyssey, the individual seeks as the Delphic Oracle inveighed, to '*know thyself*.' Such a never-ending search has, as its goal, the essential hope to distill the eternal from the transient. The religious vision embodies this realization as a "*oneness with God*." Our more secular world has yet to distill, derive, and disseminate an equivalently persuasive metaphor for the meaning of existence [10].

7.4. The *Chartres* Labyrinth

Certainly one of the largest, and sadly the only remaining authentic medieval labyrinth [11] that of *Chartres* is just over 42 ft. in diameter (12.885m). This is about as large as it is possible to be since, as a circular image, it has to fit exactly between supporting pillars of the Nave at its widest point. The survival of the labyrinth here at *Chartres* is exceptional since in the seventeenth and eighteenth centuries many of its peers were sadly destroyed. With their original meaning largely lost [12] and the then contemporary clergy unsettled by the sight of large crowds following the twisting paths, the labyrinth of *Auxerre* was destroyed in 1690. This was followed by that of *Sens* (1768), *Rheims* (1778) and *Arras* (1795). Sadly, the latter Cathedral itself was totally demolished a short time later [13]. One of the later labyrinths to be destroyed was that of *Amiens* (1825) but happily it was then reconstructed only sixty-nine years later in 1894.

As I noted earlier, *Chartres'* treasure did not escape completely unharmed and the central boss showing *"Theseus and the Minotaur"* has sadly disappeared. However, the metal studs which once attached it to the floor can still be seen today. It is the pattern of these studs which provide us with the only bare indication of how the original line engraving of *Theseus and the Monatour* might have appeared. The hypothesis that the affixing metal studs might actually represent a particular star constellation, reflected on the surface of the plaque itself, has since been discounted. One account has the plaque being removed during the French Revolution for the value of the metal as a basis for ordnance. Another suggestion is that, like the cathedral bells, it was taken to support Napoleon Bonaparte's military aspirations.

Given the witness accounts of 1792, the latter story seems to be the more supportable one. There is also a postulation that the distribution of cathedrals dedicated to the Virgin Mary across the landscape of northern France replicate the same pattern as the stars in the constellation *Virgo*. For the major cathedrals, this proposition is most probably false. However, if there are enough churches dedicated to the mother of God, as there surely are, then we can probably generate some more or less satisfactory pattern matching with the constellation's configuration. Sadly, this is simply a form of cognitive and perceptual *pareidolia*; a temptation that meets us every day in our unending search for both pattern and wonder together [14].

As well as the surrogate nature of the journey, the physical configuration of the labyrinth looked to generate the longest possible path within the outer diameter of its necessary confined space. We only have prior drawings of many of the other labyrinths, but these ubiquitously indicate a tortuous journey in which the individual finds themselves turning back upon their prior direction and thus themselves on numerous occasions. For example, the *Chartres* labyrinth's diameter is (as we have seen) approximately 42 ft. However, the patterned configuration presents a completed journey of almost 860 ft (261.5 m) in total length. This characteristic means that the physical structure, in order to present symmtericality must be very carefully conceived and fabricated. In the case of *Chartres* it involves the use of two very distinct sorts of differently colored stones to physically and symbolically juxtapose the dark and the light. It is interesting to note that in some Cathedrals the dark stones denoted the pathway; perhaps suggesting that we walk in shadow in this life. However, others like *Chartres*, have the white stones, which show the correct path; again perhaps emphasizing that the true path is one illuminated by God's light. In either case it is only the balanced Manichean contrast of good vs evil and in the form of white and black, which makes the journey possible in the first place.

7.5. Involuntary Transportation

One of the central elements I have emphasized in this present work is the process of self-generation. Indeed, I believe it is often, but not always absolutely essential to the development of the experience of wonder, that individuals embark upon this journey freely and of their own volition. Without this freedom of expression and exploration, very rarely will there be the generation of true moments of wonder. However, we must also be very aware that some, if not most, of the acts of mass human migration have not been undertaken by choice but rather have been imposed by external necessity, usually by some external authority. Whether it be the forced, mass migration of the Israelites in exile, the slaves of Africa to the Americas or the Nazi relocation of Jews from their homes to the Death and Concentration Camps of Eastern Europe, a large exodus of people often derives from forces beyond those people's own control. These forced migrations are not episodes of transcendence but rather are the pragmatic and often tragic journeys to final demise.

However, perhaps some of these nominally forced "marches" are not totally without some degree of redemption. For example, the enforced 'transportations' imposed on their own populace by the nascent British Empire of the late seventeenth and early eighteenth centuries provides us with an instructive episode. Like much of the later emigrations from central Europe, the English transportation system began first to the colonies of the Americas. Here, the English had established in-roads into and claims on native lands. What was required were individuals to populate these claimed areas. The voluntary rate of emigration, despite some rather egregious and misleading advertising, proved insufficient to the political requirements. Therefore, enforced emigration was enacted. Sadly, these emigrants who eventually became Americans proved, in the end, to be a fractious and independent bunch and subsequently the outlet to America was severed by a small altercation later termed the "*American Revolution.*" Britain still had people to export but fortunately Captain James Cook had recently "discovered" new lands to be populated, although again the indigenous 'native' population received no plebiscite on such a policy! Thus "transportation" was promoted as a criminal punishment and *Botany Bay, Australia* was the terminal destination for such putatively malfeasant individuals [15]. Transportation here was used as a device to punish many a crime that today we would barely consider as minor or inconsequential and in the form of a misdemeanor at worst. Many of those sentenced were not hardened criminals at all but rather simply ordinary individuals *in extremis* for some reason or other; most often associated with the blighted financial times and the egregious mal-distribution of wealth of that era. Evidence of the profile of relative innocence derives from the low re-offence rates in the Australian colonies where these transported individuals now possessed a first semblance of self-determination and a 'new' chance in life to succeed [16]. In light of these overall observations, I think we can conclude that we cannot "force" a sense of wonder into anyone. Mandating journeys is no way to induce any form of delightful transport. The process must be a self-generated, or at the least consensual one.

7.6. Out of the Labyrinth

For wonder, we must then really have self-determination. Choice is central to the process. I hope that this brief exposition in respect of gothic labyrinths has served to establish my point that the 'transports' I am referring to serve as a catalyst in active participants. One cannot here be a passive 'customer' and hope to achieve the same experience. It is the tragedy of our times that so much of modern technology asks, or even demands, that we be just passive observers of events conveyed to us in a second hand manner and not active participants in them. Indeed, perhaps even the church has under-estimated the need for the physicality of participation. The gradual creeping inclusion of chairs and pews into churches contrasted very much with what was, in the twelfth and thirteenth centuries, a much more physically active and participative service.

Elsewhere, I and others have argued for the critical interaction between perception and action [17]. From this psychological perspective, action can and does lead to perception just as much as perception leads to action. In all the 'transports' I have discussed to this point, a great boost is given to the inherent human locomotor capacity. In the case of liners they literally "walk on water" in the case of airships they literally "fly." But in neither of these cases is the involved individual simply or merely an armchair observer (although it is possible to *pacify* any such experience). If we hold to the earliest participatory protocols of places such as *Chartres* and *Beauvais* [16] the nominal 'passengers' were themselves largely responsible for the initiation of transportation; by their very entry into the edifice It is only through this symbiotic and reciprocal interaction of perception and action that the individual can open up the conduit of technology to free themselves from their inherent cognitive prison. Indeed, to reach a state of rational ecstasy which each transport promises, it must necessarily be an embodied experience. The full delight of cognitive orgasm can be achieved in no other way.

Reference Notes: The Riddle of the Labyrinth

[1] Solnit, R. (2000).*Wanderlust: A history of walking*. Penguin: New York.

[2] see for example: Clark, A. (1997). *Being there: Putting, brain, body and world together again*. MIT Press: Boston, MA.

[3] see Hani, J. (undated). Labyrinths. In: *Notre-Dame de Chartres*. (Translation Malcolm Miller). A labyrinth, unlike a maze is unicursal and has only one point of entry and exit and contains no dead end. There are additional definitional differences between the two. It is thought that labyrinths go back anything to 400 years ago, i.e., 2000BC, and see http://www.labyrinthos.net/

[4] Dante: Divine Comedy. See:
http://etcweb.princeton.edu/dante/index.html

[5] Bunyan, J. (1678). *Pilgrim's progress*. Nathaniel Ponder: Cornhill, London.

[6] This notion of reducing a journey to a linear map is re-captured in Ogiby's "strip maps" of late 1600's

[7] The most evocative location to experience this particular myth is at the Palace of Knossos outside Heractyion in Crete, one of the great architectural sites of the ancient world.

[8] "Labyrinths can be found in almost every religious tradition around the world. The Kabbala [Ha Qabala], or Tree of Life, found the Jewish [and other] mystical tradition[s] is an elongated labyrinth figure based on the number 11. The Hopi medicine wheel, based on the number 4, and the Man in the Maze are just two of the many Native American labyrinths.". (see: http://www.halexandria.org/dward021.htm).

[9] See: Mappa Mundi. http://en.wikipedia.org/wiki/Mappa_mundi

[10] and see Kurtz, P. (1991). *The transcendental temptation*. Prometheus Books; Buffalo, New York.

[11] Chartres contains an eleven-circuit labyrinth. This has been associated with the Knights Templar. There is an assumed association with the Temple of Jerusalem. And see Hancock, P.A. (2014). *Hoax springs eternal: The psychology of cognitive deception*, Cambridge: Cambridge University Press, see also:
http://www.labyrinthos.net/introduction1.html

[12] Even at *Chartres* one Cannon Souchet described the labyrinthine adherence as "*a senseless game and a waste of time.*"

[13] *Arras* Cathedral was sadly destroyed during the French Revolution by an individual most unaptly named Joseph LeBon. It was the hometown of Robespierre, see: http://en.wikipedia.org/wiki/Arras.

[14] Hancock, P. (2014). *Hoax springs eternal. The psychology of cognitive deception.* Cambridge: Cambridge University Press.

[15] Hughes, R. (1986). *The fatal shore: The epic of Australia's founding.* Random House: New York.

[16] see Hancock, P.A. (2009). *Mind, machine and morality.* Ashgate: Chichester. And see also: Rees, S. (2001). *The floating brothel.* Sydney, Hodder.

[17] Primarily see the works of psychologist James J. Gibson; Gibson, J.J. (1979). *The ecological approach to visual perception.* Boston: Houghton Mifflin. See also my own: Hancock, P.A. (2009). *Mind, machine and morality.* Ashgate: Chichester. as well as Hancock, P.A., Flach, J.M., Caird, J.K., & Vicente, K J. (1995). *Local applications of the ecological approach to human–machine systems, Vol. 2.* Lawrence Erlbaum Associates, Inc, Hillsdale, New Jersey

[18] Sadly, I have yet to find any reference to a labyrinth in *Beauvais*, although this is not to say there would not have been one.

"Any beautiful object, whether a living organism or any other entity composed of parts, must not only possess those parts in proper order, but its magnitude also should not be arbitrary; beauty consists in magnitude as well as order. For this reason no organism could be beautiful if it is excessively small (since observation becomes confused as it comes close to having no perceptible duration in time) or excessively large (since observation is then not simultaneous, and observers find that the sense of unity and wholeness is lost from their observation, e.g., if there were an animal a thousand miles long). So just as in the case of physical objects and living organisms, they should possess a certain magnitude, and this should be such as can readily be taken in at one view," [1]

8. SHIPS OF THE SOUL

8.1. Searching Beyond Surfaces

Having provided what are predominantly fact-based accounts of the respective 'transports' and further, having considered aspects of the 'human' and the 'technical' facets of their individual stories as well as those of their near siblings, it is now time to delve deeper and look specifically at certain physical characteristics and dimensions that they each share. Here, we must look beyond the immediate differences of stone, steel, and fabric, and identify the commonalties that these respective transports possess. The initial commonalities are those of form and function. In Table 8-1 which follows, I have identified some of these unifying elements.

© Springer International Publishing AG 2017
P. Hancock, *Transports of Delight*, DOI 10.1007/978-3-319-55248-4_8

SHIPS OF THE SOUL: AIRSHIPS, LINERS, AND
CATHEDRALS

Similarities of Form and Function

- Represented height of technology at the time of construction.

- Approximately the same height, size, and enclosed volume.

- Each contain a Staff (Captain   & Crew/Bishop & Clergy).

- Each regulated by bells.

- Congregations/passengers divided by class.

- The higher the class, the nearer to the heart of operations (i.e., altar, bridge, & fuhrergondel).

- Each are the same basic shape: Nave, Hull, Airframe.

- Each represents a hollowed out, then enclosed space.

- The altar, bridge, and fuhrergondel (i.e. control points) are each 20% the way down the structure.

- Their respective heights are almost the same but the highest point is never used by people (flag versus weather vein; each tell the direction of the wind).

- There is a manned location two-thirds up the highest point (bell tower & crow's nest) for observation and communication purposes.

- Each sub-serves transport; part physical, part spiritual.

- Each share a time sequence of process (boarding-passage-disembarkation: processional-service-recession).

- Permanent crew while passengers/congregation are temporary. (However, congregations and passengers return time and again).

- Neither passengers nor laity get to see the workings of the the transport except at special request.

- Growth pattern is the same, sudden concentrated appearance, an almost equally as sudden cessation.

- Science leads to technology leads to airships/liners. Theology leads to architecture leads to cathedral.

- Settings are similar in that they 'appear' over a horizon.

- Each approach across the surface and give the first impression of being small and close but are actually large and far away a 'visual illusion.' (Looming)

- Their physical 'scaling' against an individual is common.

- Each arose fairly suddenly, i.e., quick onset after the first prototypes were constructed.

8.2. On Common Physical Dimensions

Let us first address the similarities of structure, form, and function between these 'different' constructions. Initially, the simplest comparison concerns the physical dimensions but here we must add an important initial caveat. Recall that the "haute oeuvre" of *Beauvais* was never actually completed and so its proposed dimensionality remains not precisely defined. Sadly we can never be certain of how the final version would have appeared. To overcome this difficulty, I have substituted the physical dimensions of *Chartres* and I hope the reader will permit me this one indulgence here. From keel to mast head from ground to steeple top, from gondola wheel to apex, *Titanic*, *Chartres*, and *Hindenburg* are remarkably similar in height. *Titanic* was 175 ft. tall, *Chartres* is 167.3 ft. high and *Hindenburg* was 168 ft. from top to bottom [2]. On all of these structures, the highest point was never occupied by a human crew member. Rather this point was employed for either a flag or weather vane, each primarily used to show the direction of the wind; this being true even for *Hindenburg*. This similarity in height is indeed startling. However, if we have a rather amazing degree of agreement in one dimension, there are also strong similarities in others. Thus *Chartres* is 100 ft. wide while *Titanic* was 92.5 ft. and *Hindenburg's* diameter was 135.1 ft. But we have to look at the differences also, in this case their in length. *Hindenburg* was 804 ft. long while *Titanic* was 884 ft., but Chartres is only 426 ft., just over half the length of the other two.

Despite this difference in length, the addition of the cross transept of *Chartres* Cathedral serves to bring the absolute volume of enclosed space of all three of these transports into intriguing agreement at very close to seven million cubic feet each. The sense of mass of all these 'transports' also proves to be a common experience. Their 'scaling' against the stature of one single individual human observer is also interestingly similar. In a term proposed by the scientist James Gibson in his ecological approach to psychology, their respective perceptual 'affordances' are essentially coincident [3]; this despite their obviously different origins. Today, these examples are no longer the largest building, the largest ship, or even the biggest airborne object since each of their dimensions have been surpassed and exceeded by more modern technologies [4]. However, such apparent coincidences are not simply a reflection of my own particular personal selections process here, as we shall see.

I believe a personal story here can help to fix this notion in a more concrete manner. During a scientific conference I was attending in Norfolk, Virginia, I had the opportunity to watch the *USS Enterprise* leave harbor [5]. This first nuclear-powered aircraft carrier of the United States Navy was so vast and massive it was hard for me to actually 'see' and take in all of it in one glance. As my eyes moved over the ship, a visual illusion of the strongest kind occurred. For at that moment I was actually unable to tell whether the ship was leaving the harbor or whether the harbor was leaving the ship! Of course, my brain tried to tell my eyes that the latter eventuality was not a possible one but the momentary illusion, derived from my direct vision, was curiously and solidly convincing.

In an analogous manner, I remember taking my five year old daughter to see Howard Hughes airplane the '*Spruce Goose*'- in Long Beach California [6]. When we entered the domed enclosure, the plane was so large she could not 'see' it. She could only fixate on the full size model of the '*Wright Flyer*' which was set in front-of the vast wooden aircraft. In a childish voice, she complained that it wasn't that large and even though I tried to point out the actual '*Spruce Goose*' to her a number of times, she could never comprehend it as a single object. To this day she has no memory of that plane at all. [7] The dimensions of the transports we have examined are, proportionately, not so large. Perhaps they were so impressive precisely because their designers were careful to scale them so as not to be so big as to overwhelm perception itself!

Figure 8-1. The inside of the cathedral represents nothing so much as the upturned keel of a ship.

8.3. Hull, Nave, and Airframe

It is not merely in their respective physical dimensions that they pose intriguing similarities. All of these structures provide a vivid visual impact in terms of large enclosed spaces, as entry into any of the gothic cathedrals of Northern Europe attests. The inside surfaces are truly as grand as those of the outside; and all these structures are meant to impress. Each is also constituted of layers of technical defenses that surround and protect the enveloped individuals. As such they are very comfortable and comforting enclosures - in both a physical and a spiritual sense. Indeed, in the case of cathedrals, this protection extends even to the legal salvation of 'sanctuary' which can shelter malefactors from the forces of temporal law, if only for a limited time [8]. 'Transportation' of the sort we are exploring here also requires a degree of contemplative calm that makes invasion of these spaces by external agencies such an act of violation [9]. It is one of the reasons that stories of disaster capture our imagination. It is the sudden transition from safety to danger, the moment of terror embedded in the hours of normality. Or, as I have termed it elsewhere, "months of monotony, and milliseconds of mayhem." [10]

For another morphological similarity let us look at the picture of *Rheims* Cathedral as shown in Figure 8-1. As is clearly evident, there is here the strongest resemblance between the cathedral's nave and the upturned keel of a ship. One gets the sense that *Rheims* could well act as a wonderful dry dock for *Titanic* - given that Titanic was turned upside down! Further, one can well imagine that if the magnificent cathedral doors were enlarged and then fully opened, it might even be possible to lead in and tether *Hindenburg* inside such a wonderful hanger! As is evident from Figure 8-2, the hangers of large airships themselves resemble nothing so much as an open cathedral nave.

Figure 8-2. *The dirigible hanger at Moffett Field, California, NASA Ames Research Center. Built in the era of great airships it sits today like the lost cocoon of some long forgotten butterfly.*

❖ ❖ ❖

8.4. Locus of Control

Having looked at the fundamental shape of each of these structures, it is worth examining in more detail how the internal space of each structure is used. There is a very interesting immediate equivalence as to the point at which the 'center of control' is placed. In the *Hindenburg*, the Fuhrergondel, essentially the 'cockpit' of the craft, is 81% from the rear of the airship. For *Titanic*, the bridge is 78% of the length from the stern and for *Chartres* the high altar is 78% of the cathedral's length from its entry point via the Royal Portal. Given that each of these control areas occupies a certain length, it is fair to say that the focal point of control of each of the respective 'transports' is 80% or four/fifths along the structure of each or 20% behind the "leading" point of the structure. Why? There is no obvious reason for this commonality, except that they might potentially even have a common origin. That is, there may essentially be a 'genetic' foundation which has been passed from the older to the newer structures. Indeed, this could be in the form of "meme" made materially manifest! It is from this focal point of course that commands are issued and communications made both within and beyond the transport itself. For the passengers or congregation, this control and command location is often the focus of great mystery. Few passengers are permitted into each of these 'holy of holies,' physically separated or screened from the eyes of the everyday passenger [11]. Again, this might hark back to a common origin such as the original "*holy of holies*" in the Temple of Solomon in Jerusalem. From this observation, we might then ask what was the eventual destination of a pilgrimage once the pilgrim themselves reached Jerusalem? Could it have been this veritable holy of holies? Or were twelfth century pilgrims content to "do the sights" as so many modern tourists are?

Once in a rare while, important lay persons, especially on landmark occasions, are permitted to enter these command locations and commune with those in control, usually at the behest of some high ranking official. This periodic public access serves to dispel any sense of total 'black magic' and allows the special viewer to return to their comrades to impart the comfort and safety of the reassurances that are required. The privacy of the locus of control, however, serves to constantly remind 'mere' passengers of the difference between the arcana of professionalism of those in control and the largely impotent role that they themselves are often forced to adopt. We see the persistence of this division today in, for example, the now required closed cockpit door of the modern airliner. This unfortunate division is now occurring in ever-more different forms of both physical and spiritual transportation, as well as in many other dimensions of society which fosters ever-greater impotency. As robotics and automation continue to penetrate our everyday activities such as driving, we are likely to see a further step along the road to a complete human disenfranchisement [12].

8.5. Commander and the 'Crew'

As well certain facets of their structures, the way in which each transport is serviced and operated is also common. Each ship has a captain. Cathedrals have their chief ecclesiastical officers, and being the seat of a See this is most often a Bishop. The equivalent for our different ships are their rated captains. These respective leaders are each served by their subordinate crew, differentiated by their specific functions. Again, the differentiated functions of these 'crews' prove to be very similar. Some of the crew are given over to the physical maintenance and support of the material basis of the transport itself. Some are focused on its 'direction' and are in charge of navigation – again either spiritual or spatial. Yet others are tasked with passenger management and passenger comfort. The nearer a crew member approaches to the fundamental 'purpose' of the transport, the higher the position in the hierarchy that they occupy. Like all 'crews,' promotions are made, usually from within but also from without, emanating from other comparable 'transports.' For each full crew there are executive officers in control of day to day functions. Pursers control the financial aspects of the operation while lesser functionaries serve other moment to moment needs which in modern terms includes the now almost obligatory "gift shop." [13].

8.6. Summoned by Bells

In performing each of their respective duties and functions that require virtually continuous operations, both crewmembers and passengers need to coordinate their actions in time. For that purpose, each of these 'transports' use bells to signal watches, times, shifts, and offices [14]. In the monastic era these periods were bracketed by intervals of prayer such as *lauds*, *matins*, and *compline*. Whether watches on ships and airships are directly related to these particular religious 'offices' is difficult to establish unequivocally, but the division of the day into three parts is certainly common [15]. These observations as to common process pose a problem that is seen in evolutionary theory and that I have already raised in brief. That is, are these processes 'common' purely as a result of having evolved one from another, or do they have distinct and separate origins? It would seem fairly certain that the rules and regulations as applicable to 'ships of the air' evolved directly from 'ships of the sea.' Whether the latter derived from the procedures applicable from 'ships of the soul' is an issue to which I will return. I suspect that, in actuality, there are many common bases for each of these various lines of descent. Also, it is interesting to note that two-thirds up to the highest point is a location that can be used as a lookout for each of the transports. In cathedrals, this point is in close proximity to where the bells are located to broadcast to all 'passengers' both within and without the demesne.

The respective journey of each of these transports is bound by the same procedural sequence in time. First, there is a known and published time for 'departure.' The 'passengers' or 'congregation' are summoned by bells to advise that departure is imminent. Each transport engages in an intentional "casting off" which represents a formal beginning to their journey. This is then followed by processes of comfortable, well-recognized procedures. We have the procession, benedictions (some in physical form; some as "announcements"), the "service" itself, and finally, after completion of the 'voyage' there is a formal disembarkation and a final procession in which passengers are very definitely shepherded out from the transport, albeit in a kind and caring way. The same sequence happens in virtually all auditoria and stadia when we witness, for example, sporting events and rock concerts whose function also is to suspend everyday concerns if only for a short while and permit us the opportunity for 'release.' The venue is then closed and cleaned by the maintenance personnel and readied for its next service. You can witness the forcible expulsion for yourself if you try to stay on in the arena, thus making the acquaintance of the ubiquitous 'security,' personal. It is evident from these observations that the passengers are only temporary occupants while the 'crew' can well devote their whole lives to these respective environments. Sometimes, when individual passengers return frequently, they can even overtake new crew-embers in terms of 'seniority' and come to feel they share much of the mystique of the transport. For *Titanic* and *Hindenburg* few such affiliations were able to be generated because of their short–lived existence. For the cathedral of *Beauvais*, as with more complete cathedrals, such allegiance can run for a lifetime (and assumedly beyond?) and this is as true today as it was in former times.

8.7. Congregated 'Passengers'

Although possessing a captain and crew, the major raison d'etre of each one of these respective transports is fundamentally to serve its passengers. However, all passengers were not and are not equal! For each transport, passengers were fundamentally divided into their various 'classes,' although there were so few passengers on the *Hindenburg* there were only very subtle variations in class here. The richer the passenger, the closer to the locus of control and the better view they had of the sights to be seen [16]. Selected high-class passengers secured privileges such as seeing the locus of control, sitting at the Captain's Table, etc. Moving from one's assigned location proves problematic for all of these transports. The movie *Titanic* and Leonardo DiCaprio's character notwithstanding, mobility between the assigned 'class' areas was highly restrictive and remains so even on today's modern transports. It is a further similarity that passengers who return time and time again tend to gain privilege. Thus, as they become a recurring part of the transport, this semi-permanence is recognized, acknowledged, rewarded, and even then compared to the permanence of the crew. Returning 'passengers' would be recognized and greeted but their presence was always necessarily transient compared to the omnipresence of the crew. Although this is common across the 'transports' I have discussed, it is in fact a more ubiquitous distinction between the server and the served, the passenger and the crew, the professional and the amateur, the active participant and the passive recipient. When we later consider purpose, this difference becomes vital.

8.8. Looming and Awe

To this point, I have looked almost exclusively at the structures themselves but it also instructive to look at each of them within their setting or, more generally, their context of operations. Virtually all cathedrals are sited to provide their viewer's a physical sense of awe. It is not only the example of *Chartres* as we have seen but with many others such as *Lincoln* and *Durham* which are built on prominent locations especially to tower above their landscape and emphasize the majesty of their appearance simply by their position (see Figure 8-3). Whether on a ridge or a promontory, almost all cathedrals prove impressive as part of the very siting. Ocean liners also tend to dominate their surroundings in almost all circumstances. Indeed one rare exception to this dominance comes when they are docked with sister ships in major harbors like New York. Here, I have illustrated such an event in Figure 8-4. In the normal run of things, they dwarf everyday ocean waves on the vertical scale but are themselves dwarfed by the ocean on a lateral scale. *Hindenburg* simply appeared and is probably still the largest flying object with the specific purpose of transporting commercial passengers to have ever been created. Even though seen only periodically, *Hindenburg* caused a sensation whenever she arrived over a location. For example, recall the stopped baseball game and the accounts of traffic halting in the streets of New York, Boston, and other locations in general. Little wonder that creations five times as large as *Hindenburg* can be mistaken for UFO's especially when seen against the ganzfeld of a featureless sky [17].

Figure 8-3. *Illustration of Lincoln Cathedral showing its dominant position even today above the buildings of the contemporary city. We might well imagine how much more impressive this would have been at its inauguration.*

Sadly, I have only one comparable personal story here, since I never personally witnessed *Hindenburg* or any of her peers in the air. Upon our recent arrival in central Florida, my family and I were off for an evening excursion heading west along one of the major freeways. Suddenly, cars started to pull over to the side of the road for no apparent reason. We followed suit, thinking there was some obstruction further ahead of us. We were wrong! After parking, we left the car and turning round saw a splendid sight. There, in the twilight of evening was the space shuttle launching from Kennedy Space Center. It made its spectacular way into the heavens leaving behind a contrail of many colors. Much more of the nature of the 'fiery messenger' than the quiet stately progress of *Hindenburg* nevertheless this was yet another instance of a *'transport of delight.'* Indeed, as I will argue later, space exploration is the transport of delight of our times and the pity is that, as yet, so few can personally partake and share in the experience [18]. For all the structures we have been discussing, they appear, at first, to be small and paradoxically relatively near at hand. However, they are exceptions when set against a normal context, and so each presents an artificial visual illusion. The sudden realization of this illusion and its dispelling by the mind is a critical first recognition of the extraordinary. It certainly was for me.

Figure 8-4: *A rare moment in time when three leviathans of the sea docked together. There is even space for Titanic!*

There is a scene in James Cameron's visually marvelous film *Titanic*, in which the star of the film makes her first entrance, rather like Rita Hayworth's making her first appearance in *Gilda*. For, the true star of Cameron's movie is, of course the ship herself. Cameron's vision of this event is typically acute. He uses one of his human stars, Kate Winslett (playing the role of Rose de Witt Buchcater) to emphasize this critical effect. If you have seen the movie you will already know the scene of which I am speaking. If you have not seen the movie, the scene is fortunately easy to describe and now access. We see Rose, the spoilt teenager turn suddenly to the ship and in true character she remains childishly and even petulantly unimpressed by this marvelous vessel.

As viewers, however, we can only catch our breath at this first view of the ship of the age. Why is this? Why are we not indifferent like the palled and pallid Rose? I want to suggest one reason for this sense of awe; although I hasten to add that I do not think it is the only reason. I believe this effect is due in part to a perceptual process which is called 'looming.' What is "looming?" Although it has a complex scientific definition, looming is very much what it sounds like. Normally, when we think of something as looming over us, we might think of Dracula spreading his dark and voluminous cape and looming over his latest hapless, female victim. In fact, the cape here proves to be absolutely essential. Looming occurs when an object or entity occupies a large portion of the persons' visual field. You can try this experiment for yourself very simply. Most of the time our visual field is cluttered up with any number of discrete, individual objects of different sizes at different distances and so no one particular thing dominates all of the others. However, if you put some dark object in the world and have another person bring it slowly toward you, it will at some point trigger a fear response as your perceptual system warns you of an impending collision. William Schiff, who did extensive work on this issue, called this looming process *'optically explosive'* [19]. It is a very perceptive observation.

You can still derive looming effects from a relatively small object but they have to be very close to your eyes. However, you can get a much more impressive looming effect from a large object that is further away but approaching quickly. What I am claiming here is that one characteristic of each of the transports we have been discussing is that, in their presence, the individual is dwarfed and so gets part of their sense of awe, from their own, self-generated recognition of just how small they are! The same sense accompanies anyone who can, with knowledge, look up into the night sky. [20]

8.9. The Common Ultimate Function

Ultimately, the purpose of each of these technologies was to 'transport' their patrons to other realms. Whether that other realm was a spiritual one or a physical one depends largely upon one's perspective. Cathedrals appear to be designed almost exclusively for spiritual transformation [21]. However, we must remember that when they were built they also represented radical physical transformations from the hurly-burly of the outer world to the quiet contemplation of the inner cavern of the gothic nave. Anyone seeking a similar transformation today only has to visit *Lourdes* where the transition from the rabid commercialism of the town to the quiet peace of the religious site via a bridge across a dividing river is as stark as any I have ever experienced. Entry into a large cathedral, leaving behind the bustle of a major city provides a very similar effect. In accord with this notion, liners and airships look superficially as though their only purpose is the physical translocation of people. However, only scratch the surface and we find that many of the 'passengers' secured their respective "tickets" in the hope of a better life. Thus, their vision was one of a better life in a land of ultimate promise [22]. Whether this was "heaven" for nineteenth century immigrants is moot, the proposition is just the same. What was on offer was a promise of change, a promise of something better, a better life for self and family and possibly a glimpse of heaven, the chance of a new start. Such transports represent our hopes made manifest that technology can take us beyond where we are, beyond ourselves and as 'transports of delight' deliver us into a divine realm of eternal happiness.

❖❖❖

Reference Notes: Ships of the Soul

[1] *Aristotle, Poetics, c. 326BC.*

[2] Each of these dimensions can be found at standard sites such as:
http://www.airships.net/hindenburg/size-speed.
http://en.wikipedia.org/wiki/RMS_Titanic
http://en.chartressecrets.org/cathedral/background_history.htm

[3] Gibson, J.J. (1979). *The ecological approach to visual perception.* Boston: Houghton-Mifflin.

[4] Hancock, P.A. (2001). Five times the Hindenburg! (letter). *Skeptical Inquirer,* 25 (4), 76-77.

[5] http://nation.time.com/2013/01/07/obit-for-a-carrier/

[6] see: Winchester, J. (2005). *Hughes H-4 'Spruce Goose'." Concept Aircraft: Prototypes, X-Planes and Experimental Aircraft.* Kent, UK: Grange Books. And see: http://evergreenmuseum.org/the-museum/aircraft-exhibits/the-spruce-goose/

[7] The Wright Flyer actually flew a distance shorter than the Spruce Goose is long.

[8] Shoemaker, K. (2010). *Sanctuary and crime in the middle ages.* Fordham University Press. Bronx: New York.

[9] Such violations have been relatively infrequent as far as "sanctuary" is concerned that when they occur, such as Edward IV's entry into *Tewkesbury Abbey* at the *Battle of Tewkesbury* in May 1471, the discussions and repercussions are extensive.

[10] Hancock, P.A., & Krueger, G.P. (2010). *Hours of boredom – moments of terror: Temporal desynchrony in military and security force operations.*

Center for Technology and National Security Policy: Defense and Technology Paper.

http://www.ndu.edu/ctnsp/publications.html

[11] http://templemountfaithful.org/articles/temple-location.php

[12] see: Hancock, P.A., Billings, D.R., & Oleson, K.E. (2011). Can you trust your robot? *Ergonomics in Design, 19* (3), 24-29, and: Hancock, P.A., Billings, D.R., Olsen, K., Chen, J.Y.C., de Visser, E.J., & Parasuraman, R. (2011). A meta-analysis of factors impacting trust in human-robot interaction. *Human Factors, 53* (5), 517-527.

[13] see e.g., Barnes, C.F, Jr. (1938). Cathedral. In: Joseph Strayer, (Ed.) *Dictionary of the Middle Ages.* Vol. III. (pp. 191–192), New York: Scribner's.

[14] Betjeman, J. (1989). *Summoned by bells.* John Murray: London.

[15] http://www.yale.edu/adhoc/research_resources/liturgy/hours.html.

[16] It is nice to believe that such differentiations of class actually generated its own vocabulary, such as the 'U' 'Non-U' differentiation of Nancy Mitford (1956). (Ed.). *Noblesse oblige: An enquiry into the identifiable characteristics of the English aristocracy.* London: Hamish Hamilton. The epitome is often held out to be travel to and from India in which the shaded side of the ship was port-out and starboard home, or the POSH side. Sadly, this may well be a myth, although still an attractive one: Quinion, M. (2004). *Port Out, Starboard Home: And Other Language Myths.* Penguin: London.

[17] Hancock, P.A. (2001). Five times the Hindenburg!. *Skeptical Inquirer, 25* (4), 76-77.

[18] It is sad to report that just as I was completing this work there was a tragic crash of the Virgin Galactic space vehicle one of whose avowed purposes was to help promote general passenger travel into space. And see: en.wikipedia.org/wiki/VSS_Enterprise_crash.

[19] see for example: Schiff, W., Caviness, J.A., & Gibson, J.J. (1962). Persistent fear responses in Rhesus Monkeys to the optical stimulus of "looming. *Science, 136* (3520), 982-983. And see also: Hancock, P.A., & Manser, M.P. (1997). Time-to-contact: More than tau alone. *Ecological Psychology, 9* (4), 265-297. Such looming has been shown, in and of itself, to capture human attention, see: Yantis, S., & Jonides, J. (1984). Abrupt visual onsets and selective attention: Evidence from visual search. *Journal of Experimental Psychology: Human Perception and Performance, 10,* 601-621.

[20] *During the day you can see a hundred miles, but at night you can see a billion years* - Hancock.

[21] See: Erlande-Brandenburg, A. (1995). *The Cathedral builders of the middle ages.* Thames & Hudson: London.

[22] and see Hancock, P.A. (2009). *Mind, machine, and morality.* Ashgate Publishing, Aldershot, England.

"[The R101] is as safe as a house-except for the millionth chance. [1]

9. THREADS THROUGH TIME

9.1. Tying Things Together

At this juncture, I want to elaborate on one particular thread
through time which serves to link these various transports directly
together. This connection emanates, not from the German airship
industry but rather from the English efforts to keep pace with, or
even outpace, their Teutonic rivals. In the early 1920's England
aspired to inter-continental air travel through the construction of
two of the then largest rigid airships ever to be built. The hope
was to tie together the far-flung outposts of the British Empire
with the mother country. In particular, the aspiration was to link
India, Australia, and Canada closer together with England as the
focal point. (I cannot pass up on this observation without noting
how similar this aspiration was to some of the reasons advanced
for the construction of the two Cunard Liners *Mauritania* and
Lusitania some decades earlier). It was, of course, these very
Cunarders who were the direct stimulus for the White Star
response with *Titanic, Olympic,* and *Britannic*.

The two British airships so conceived were designated *R-100*
and *R-101* respectively. They were to be built in two rather
differing ways [2]. The *R-101* was to be fabricated by the
government (and was accordingly labelled the 'socialist' airship).
In contrast, the *R-100* (the so-called 'capitalist' airship) was to be
built by private industry. It is true that the government actually
financially underwrote each of these projects as well as

© Springer International Publishing AG 2017
P. Hancock, *Transports of Delight*, DOI 10.1007/978-3-319-55248-4_9

shouldering the cost of the necessary ground stations and infrastructure at Cardington in England, at Ismailia in Egypt, at Karachi in India, and finally at Montreal in Canada. Had both airships been successful they would have truly revolutionized travel across the British Empire and, by extension, across the globe itself. Sadly, as we shall see, this admirable and vaunting dream was never to be realized.

9.2. Airship and Cathedral

Problems dogged the manufacture of *R-101* from its very beginnings. The insistence on diesel engines that were under-powered and overweight drastically reduced both R-101's vital lifting capacity as well as her maneuvering abilities. Also the internal architecture of *R-101* left it vulnerable to an inadvertent deflation of the gas volume. This combination of greater weight with lower lifting capacity was to be a critical factor in her existence but what eventually proved perhaps most problematic was a last-minute design change that saw an additional 'section' of 46 ft. added to the center of the airship. This insertion was ostensibly to compensate for her now disappointing overall lifting capacity. The addition brought *R-101* to a final length of 731 ft. [3]. At the time of her launch in 1929 she was the largest flying craft ever made by human hands. Indeed, it was not until *Hindenburg* herself was launched some seven years later that *R-101's* record was eclipsed [4].

During construction, political pressures has been mounting on Lord Thompson, the Air Minister. He certainly needed the boost of a successful flight, having tied his personal reputation to the project. Indeed, his full personal title was '*Lord Thompson of Cardington*' which, as we have heard, was the Bedfordshire home base of the two airships. In contrast to its socialist peer, the commercial *R-100* had already completed a successful trans-Atlantic flight to Canada. The *R-101* team were keen to at least match if not surpass the achievements of her commercial sister ship. Such diverse social and technical pressures epitomize the growing precursors to disaster which were now beginning to coalesce [5]. More hold-ups were experienced when the airship's fabric covering turned out to be flawed. This delay saw even greater pressure building on all of those associated with the airship. Eventually *R-101* took off, literally with a last-minute air-worthiness certificate, the fulfillment of which was dependent upon her performance on her maiden voyage. This was to be the record breaking outbound trip to India [6]. It was the true tragedy of the *R-101* therefore, that she was completely aerated before she was fully air-rated (see Figure 9-1). When she started her maiden voyage on the evening of 4th October, 1930, there were on-board her designer, V.C. Richmond, as well as numerous VIP's including Sir Sefton Brancker the *Director of Civil Aviation,* as well as *His Majesty's Secretary of State for Air*, Lord Thompson of Cardington himself. Operating the airship was a crew of 42 and all told, there were 54 souls on board [7].

Figure 9-1: *The original airworthiness certificate of R-101.*

The last link in *R-101's* particular, idiographic chain of disaster was the weather. Dark and stormy in England, the winds were projected to be gusting to over 50 mph in northern France along the immediate track of the airship. Rocking to and fro longitudinally in these adverse conditions, *R-101* did begin to lose gas and subsequent lift from what was already too small a margin for error. This increasing risk was present even before she crossed the southern English coast. But added to these issues, her rain soaked cover weighed the ship down even further; now to crisis level. When their revised course took them toward a ridge which was notorious for its turbulent winds, their fate was, to all intents and purposes, sealed [8].

At just after 2:00 am on October 5th shortly following a watch change, the ship went into a terminal dive. Unable to release ballast from the control car, the officer of the watch could only order an emergency climb and it was in this 'stalled' mode that *R-101* first hit. One crew member jumped from an engine gondola as the ship subsequently 'bounced' some 60 ft. back into the air. The next impact was her last as she broke her back at exactly the point where the controversial new section had been inserted. The formal inquiry however identified a probable tear in the outer cover as the major proximal contribution to her final, fatal, and fiery failure [9]. The gas then ignited and forty-eight individuals died in the early hours of that October day. This left only six people alive to tell the tale. Sadly, two of these nominal survivors also shortly succumbed to their injuries. The total number of fatalities on that October night was greater than that of even the subsequent *Hindenburg* conflagration. In doing so, it proved to be the second greatest ever airship disaster in terms of loss of life [10]. That town in northern France outside which *R-101* crashed - *Beauvais*!

The demise of the *R-101* was not destined to be a solo one. Despite being different in design and provably more reliable in performance, the *R-100* was at first grounded and then dismantled by an embarrassed government which had been shown up by a commercial organization as to how to create a cheaper and more reliable craft [11]. Many individuals proved to be understandably bitter at this decision. However, it has subsequently been observed that the *R-100* was perhaps purposely designed in a more conservative manner while *R-101* was, in contrast, purpose-created to *"push the envelope"* of technology to a much greater extent. Barnes-Wallis, the subsequent inventor of the war-time 'bouncing bomb' [12] and the designer, architect, and mastermind behind *R-100* was much disillusioned by this exhibition of governmental perfidy; he should be alive today!

An influential biography from the pen of the talented novelist Nevil Shute also served to exacerbate the dispute and recriminations from the perspective of the *R-100* group [13]. Shute had been part of the *R-100* team in his professional capacity as an engineer. The crash of the *R-101* proved to be the end of Britain's formal flirtation with lighter than air craft. The final termination of this technology in flames was, of course, experienced in the subsequent *Hindenburg* disaster seven years later. That event ended the whole technology; at least to our present time.

9.3. Recycle and Resurrect

R-101's main structure was composed of a framework of steel and duralumin. The latter was a rare and expensive material. After the crash this valuable wreckage lay all over the French countryside, framed against the background of her eight hundred year old stunted cathedral counterpart. Both technologies had aspired to, and had reached out, into unprecedented territory. However, each had gone just beyond the edge of stability. The grand airship had been destroyed and all that was left was her charred frame, Figure 9-2. The vision for this technology had been to straddle the Earth, to link together the fulcra of Empire. Now the vision lay burned, charred, and dowsed; straddling only a very small part of the Earth and the grave to forty-eight souls. One 'cathedral' had crashed, almost into another. However, like a

Figure 9-2: *The wreckage of the R-101 lies strewn across the countryside of northern France. Of the 54 people on board, only six survived.*

Figure 9-3: *The rear view of the crashed airship.*

mythical phoenix of old part of *R-101* arose again. For, the
wreckage of the *R-101* was acquired privately by Thomas Ward
Ltd of Sheffield, England. However, five tons of the valuable
duralumin was subsequently acquired by the Luftschiffbrau
Company to be recycled and used to help in the building of their
latest creation [14]. This latter design was known as the *LZ-129*
project, or as we know it today – *Hindenburg.* So from the wreck
of *R-101*, blazing in the skies against the more permanent tragedy
of *Beauvais Cathedral* we find part of the material genesis of
Hindenburg. As with each of these physical incarnations, ideas
underlying all of human technical progress are ever-recycled. Of
course, the same might be said of the whole panoply of human
aspirations.

Figure 9-4: The memorial to the disaster on the roundabout on the edge of Beauvais. (Photograph by the Author).

The wreckage of the *Hindenburg* itself was subsequently reclaimed by the Luftschiffbrau [15]. However, this was not before many stalwart citizens of Lakehurst and personnel from the Naval Air-Station (NAS) itself had managed to secure some pieces for posterity. Indeed, parts of the *Hindenburg* are periodically available today on *eBay* – as I know well since this was the avenue I used to secure my own personal memento. The *R-101*, the English 'Cathedral of the Air', died within sight of its French companion, only to become part of its German progeny – the most famous airship ever created. This might well be the very emblem of the never-ending cycle of dream, vision, fabrication, realization, destruction, and finally resurrection. Is this the cycle of all life whether for living or for non-living entities [16]?

9.4. Modern Reflection, Modern Purpose, Same Goal

Instead of tracing forward from *Beauvais* to *R-101* and then to *Hindenburg* let us consider an alternative path. As we have seen, *Beauvais* was essentially the final despairing reach for height and illumination and the final resting place of England's "lighter than air" dreams. Although the collapse of 1284 curtailed the physical ascension of *Beauvais*, it never stunted the vision that it embraced. It is possible that the vision of *Beauvais* itself could be resurrected and indeed I would suggest that it has been. In Figure 9-6, we can see that *Beauvais* has not been resurrected as a cathedral of religion but rather this time as a cathedral of science! This cathedral is represented by Fermilab located just outside of Chicago, Illinois, in the United States [17]. Specifically, what *Beauvais* inspired was the Robert Rathburn Wilson Hall, the Laboratory's headquarters building. The interior is a true homage to *Beauvais*. Since all technical transports seek to 'open' passage to new 'dimensions,' the link here to one of the world's leading particle research facilities is perhaps more than pure happenstance? Not as a gateway to spiritual enlightenment directly, but as a gateway to the sub-atomic realm. Fermilab here proves to be a *'transport of delight'* to voyaging scientists! Perhaps it is also their own special gateway to spiritual enlightenment?

Figure 9-5: *The memorial in a larger context (Photograph from the road).*

Figure 9-6: *Wilson Hall, the central facility of Fermilab, the physics research facility is modeled on Beauvais Cathedral. The 'transport' to the realm of sub-atomic physics might well be physical but for many it is also spiritual — at least in a scientific sense.*

However, it is not simply science that recapitulates *Beauvais* since its further aspirations and elaborations are located in the tragic and heroic city of New Orleans, Louisiana. Here we find yet another incarnation of the lost cathedral. In the cemetery of St. Roch is a representation of *Beauvais*, but this time a scaled-down version [18]. Being lower than sea-level, New Orleans possesses a number of very interesting cemeteries in which bodies are interred in above ground vaults. This provides these respective cemeteries with their unusual and almost unique character. The St. Roch Chapel, used for final services, provides a portal to the eternal. It is an example how certain forms of dream and vision are heavily embedded in the human narrative and necessarily recur in each succeeding generation [19]. These two instances can neither be a complete nor exhaustive list as to how these ideas, notions, and representations are recalled and resurrected around the globe. But for the purpose of my argument, such examples need not necessarily be a full and complete listing. Only this limited listing is sufficient in order to substantiate the general principle. Inspiration and vision do not die, they recur with each generation and with each new technological incarnation of those succeeding generations.

9.5. Toward a Conclusion

It could appear to an assiduous reader that I intentionally pre-selected my examples of such transports specifically in order that I might now "reveal" these respective connections. However, I have to say that this was not the case. Rather, like many other aspects of history that I deal with [20], I made such connections only as the search itself progressed. Indeed, I think I can safely say that it is the search for these material and more spiritual connections that actually drives my work forward. I believe that 'connections' can be discovered and even created across almost all phenomena, the central question being how 'meaningful' such connections actually are. So, while each story is necessary to my respective thesis, such stories by themselves are not sufficient to illustrate the 'higher' principles that I look to affirm. In more specific terms, each story and the principles it illuminates must add to the overall picture but they cannot, individually, be the full picture in and of themselves. In light of this observation, I now move to a more specific focus on those higher principles.

Figure 9-5: St. Roch Cemetery, New Orleans, Louisiana.

❖❖❖

Research Notes: Thread through Time

[1] Quotation from Lord Thompson. This observation was subsequently used by Leasor as the title of their text: Leasor, J. (1957). *The millionth chance: The story of the R-101*. Reynals & Co.: New York.

[2] For further information on R100 and R101 see: Higham, R. (1961). *The British rigid airship, 1908-1931*. London: Foulis, and also: Masefield, P. (1982). *To ride the storm*. London: William Kimber.

[3] http://en.wikipedia.org/wiki/R101.

[4] http://www.airships.net/hindenburg/size-speed

[5] For a technical approach to the notion of systems-based error see: Reason, J. (1990). *Human error*. Cambridge: Cambridge University Press, and: Reason, J. (1997). *Managing the risks of organizational accidents*. Ashgate: Aldershot.

[6] The actual airworthiness certificate is illustrated in Figure .

[7] http://www.airshipsonline.com/airships/r101/

[8] It is, of course, of more than passing interest that one account of the destruction of *Beauvais* Cathedral included the suggestion of malevolent winds from England. It is more than reciprocal that the English Cathedral of the Air, *R-101*, succumbed to malevolent winds from France.

[9] http://lordkingsnorton.cranfield.ac.uk/archive%20docs/r101.pdf.

[10] In relative terms, *R-101* was the single greatest loss of civilian life in such an accident. However, in absolute terms the greatest ever loss was the Akron in which some 73 died and only 3 survived. However, it must be noted that Akron crashed at sea and it was drownings and hypothermia, like in Titanic's case, that took most individuals here. And

see: Smith, R.K. (1965). *The airships Akron & Macon (Flying Aircraft Carriers of the United States Navy)*. United States Naval Institute: Annapolis, Maryland.

[11] Higham, R. (1961). *The British rigid airship, 1908-1931*. London: Foulis.

[12] Morpurgo, J.E. (1972). *Barnes Wallis: A biography*. London: Longman.

[13] Shute, N. (1954). *Slide rule: Autobiography of an engineer*. London: William Heinemann Ltd.

[14] See: http://cardington.weebly.com/r101.html. "*Zeppelin are on record at buying scrap duralumin from the wreckage of the R101. This has never been confirmed but that metal could have gone into the building of the Hindenburg.*"

[15] See: http://www.roadsideamerica.com/story/2160, and: http://www.nlhs.com/hindenburg-lz-129.html.

 [16] Russell, B. (1915). On the experience of time. *Monist, 25* (2), 212-233.

[17] https://www.fnal.gov. And see the specifics of Wilson Hall: https://www.fnal.gov/pub/about/campus/wilsonhall.html.

[18] St. Roch is one of several of the above ground cemeteries of New Orleans. http://morbidanatomy.blogspot.com/2009/05/st-roch-cemetery-and-chapel-new-orelans.html.

[19] Pears, I. (1999). *An instance of the fingerpost*. Berkeley Publishing: New York.

[20] See: Hancock, P.A. (2009). *Richard III and the murder in the tower*. History Press, Stroud: England.

❖❖❖

"Possible, it is beyond human art to convey the sense of something lost, but eternally present, that such places inspire. In every light, and in every season, it possesses a transcendent beauty; but in summer it is very paradise" [1]

10. TRANSPORTS OF DELIGHT

10.1. Further, Ever Further

It is here in my penultimate chapter, that I now need to go beyond all individual stories and beyond the simple links founded upon physical comparisons or even the intriguing spatial and temporal conjunctions. And with this step, these final two chapters are each directly concerned with the *purpose* of all transports. I want to start by making what, at first, may seem to be a rather strange assertion. I want to claim that the fundamental purpose of each and all these diverse forms of 'transport' is *exactly the same*. Since this proposition is probably most difficult to support with respect to the gothic cathedrals, it is with them that I start.

When I was disoriented in time during that very early morning visit to *Rheims Cathedral*, now so long ago, it was as the pipe-organ began to sound that I knew that what I was looking at *was* the height of technology. I understood that it was only my own transience that rendered such a vision displaced in time. *Rheims* was of course the height of technology when first built. But there again so are all technologies when they are first created. My misperception here was due to the fact that this gothic technology had been just too good. It has lasted so long that, as a fleeting being, limited to my own narrow pocket of time, I could not "see"

© Springer International Publishing AG 2017
P. Hancock, *Transports of Delight*, DOI 10.1007/978-3-319-55248-4_10

Rheims as it truly was. I was as constrained in my own way as my daughter had been in trying to "see" the Spruce Goose. It is as if one of our own contemporary supercomputers had somehow managed to last a thousand years and was still functioning, at least to some degree. It would seem to my millennial cousin just as anachronistic then as the cathedral now appeared to me. Indeed, with the rate of accelerating technology it would probably appear to be much more so [2], it being rather unlikely that almost anything we build today will last a thousand years. An empathic understanding of this proposition only serves to show how sadly limited is the window of time through which any one of us can ever look [3].

I know now that the builders of *Rheims* and its various gothic peers were not simply reaching toward God. Their fundamental purpose was to carry all of us up with them. For both the 'crew' and the 'passengers' engaged upon this voyage, transportation to another realm was, and remains, the primary goal. The destination is always a desired but chimerical 'promised land.' Even as the congregation and clerics raise their eyes to heaven, they can never avoid the metaphor of transport. For the internal construction of all gothic cathedrals resembles nothing so much as the inside of an up-turned liner. Cathedrals and ocean liners are womb-like in their configuration and airships even more so since you are privileged to actually 'ride' inside them. You can yourself experience this transition simply by looking up at the inside any of the great gothic cathedrals of northern Europe [4]. Not merely transports then, each of these conveyances actually recapitulate your own mother! They are warm and nurturing inside; they move you and you need only exert little effort to achieve such passage; they sustain you materially and spiritually. No wonder they represent special human places and no wonder so many are dedicated to the Mother of God [5].

When these cathedrals were first built, long-distance travel, as we know it, was almost unknown. It was only with pilgrimages and the advent of the crusades at around this era of the twelfth and thirteenth centuries that such travel became an activity pursued beyond the very few exceptions such as merchants and explorers [6]. For the everyday individual of early medieval Europe, travel beyond their own small local area might even be a once in a lifetime event. And at the point in time when these cathedrals were erected, such a journey was very likely to be some form of pilgrimage to a site of sacred purpose [7]. Used to their own small places of worship and single-story wattle-and-daub churches, newly constructed cathedrals must have appeared to this sheltered and constrained populace as the most marvelous creations in the world. These buildings were truly the new *"Arks of God."* They were the veritable portal to the world beyond; the world eternal. Thus, I claim that the fundamental purpose of gothic cathedrals was indeed to 'transport' those who saw them and entered into them to another realm. That realm itself was redolent with the promise of a better life, even of perfection. In a true sense, even though cathedrals have never physically moved, these marvelous places of worship are perhaps well-described as the ultimate form of 'transport.'

From the foregoing argument, the *Titanic* might initially be seen as sub-serving a very different; and much more prosaic function. On the surface, *Titanic* was sold as the most luxurious form of trans-Atlantic passage available. However, *Titanic's* major function was, as with all of the liners of that period, the mass transport of emigrants from Europe to become immigrants to America [8]. *Parenthetically, one might be forced to wonder precisely where in the trans-Atlantic voyage did emigrant become immigrant?* Regardless of this exact longitude, it is in the nature of transition that I claim that the functions are co-incident. Cathedral and liner each represented the height of technology at the time they were first created. They were the newest, the brightest, and the most marvelous but their fundamental function was each to support human dreams. In the case of hundreds of thousands of emigrant Europeans, the dream was the hope of a better life in America and it drew them on [9]. *Titanic* was indeed a veritable *'transport of delight'* but it was truly immortalized as a *'ship of dreams.'* The dreams for most on board lay in a change in life, not merely the luxurious crossing itself; as it was for the favored few. One might even be tempted to believe that the hope of the least of the third class steerage passenger was in reality better than the jaded palate of each saturated first-class passenger. But first, one would be best advised to experience both! However, I argue that every gothic cathedral, but especially *Beauvais,* can equally be described just as accurately as a *'ship of dreams.'*

Hindenburg is admittedly, a somewhat varying case; but not sufficiently different to fracture the similarity of fundamental purpose than I have proposed. As a result of the limitations to lifting capacity, weight was an all-important issue in rigid airship flight. Even with all her innovations and new technologies and materials, *Hindenburg* was still restricted to a total lifting capacity of 511,500 lbs; approximately 228 tons. The margin of lift above the weight of the structure itself was some 22,000 lbs or approximately 10 tons. This extra capacity was used for passengers, crew, and baggage [10]. As a result, in its early configuration as a form of air transport, there were no places for steerage passengers on *Hindenburg* or any of its peers. However, I would claim that this was because *Hindenburg* was still very much a part of a highly experimental technology. And sadly of course, it was technology that essentially disappeared with her destruction.

As the evidence of Herb Morrison's presence showed, *Hindenburg* was part of the greater struggle against time which continues to this day in the leitmotif; faster [11]. At Lakehurst, *Hindenburg* was to be met by an American Airlines aircraft that would take passengers straight on to Chicago. In this sense, *Hindenburg* stood in somewhat similar a role as that of a Boeing 747 to a trans-continental supersonic *Concorde,* where the latter has, of course, now itself 'disappeared'. Passengers paid a premium for time but the 'transport' is still the same. Had *Hindenburg* completed her travel season uneventfully, we may well have seen even larger air ships trying to perhaps capture a greater segment of the trans-Atlantic trade. That she and her various relations and peers were outdone by heavier-than-air flight capability is another story of technological super-position but the accounts of those who traveled on *Hindenburg* reinforce its unprecedented sense of majesty [12]. The advertising brochure put out by the company itself rhapsodized:

> "*There is no noise beyond the distant murmur of the engines and the sigh of the wind on the outer hull. No dust, no soot to trouble you, the whole atmosphere is one of tranquility and peace. The air is delicious and fresh, in fact you seem to have been transported into another and more beautiful world.*" [13]

Beyond this self-aggrandizement, *Hindenburg* was at the cutting edge and she most certainly then represented a quintessential 'transport of delight.'

Figure 10-1. *Airship Travel Made Easy. Brochure of the Zeppelin Rederei.*

Ultimately, each of these technologies sought to translate their passengers perception of their world. Whether that realm was a spiritual one, as in the case of gothic cathedrals, or was the vision of a new life in a land of promise, as was America for the immigrants, the proposition was the same. It was the hope for change, a promise of something better, a new life, a glimpse of heaven and the chance of a new beginning. That each of the respective technologies in their aspiration to become bigger, better, and then to be the very best, eventually fell to the forces of nature is no critical reflection upon the expressed hope itself [14]; although it is most certainly a very real reflection on human hubris.

Figure 10-2. *Hindenburg over lower Manhattan, New York*

10.2. The Familiar Form of Failure

But how did these respective technologies fail? And why did they fail? There is much more to the nature of their failure than the simple descriptive accounts of individual episodes alone; evocative though these may be. I can illustrate this greater pattern, by plotting the growing physical dimensions of these respective technologies and then showing the point at which disaster occurred. In Figures 10-3 to 10-5 I have illustrated these respective progressions toward failure of cathedral, liner, and airship, sequentially. Perhaps the most illustrative is the first case. Here we can use the search for height as the proxy for technical aspiration as it moves toward incipient failure. Confining this argument for convenience to French cathedrals alone, we can see their progressive development, in Figure 10-3. This curve shows the form of an increasing exponential. That is, not only does the height grow, but it grows at an accelerating rate.

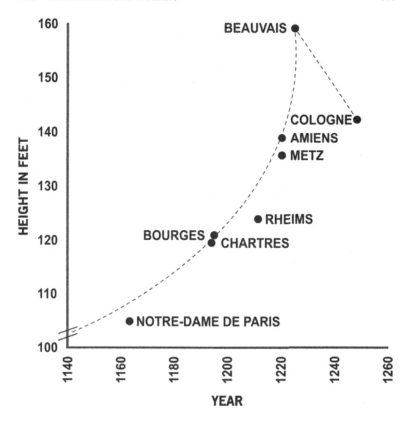

Figure 10-3: *The growth of cathedral height by year*

At first, the steps between cathedral heights (i.e., from *Paris* to *Chartres* and then from *Chartres* to *Rheims*) are relatively consistent. In this case it appears that there is a linear increase of approximately five feet in height for each of those sequential steps. But note here the different time scales involved. That second, equivalent increase took only twenty years to achieve, as opposed to the thirty years it had taken for the first five foot increment. But then the jumps in height get larger while the time interval to achieve each increment grows smaller [15]. This technology is growing here, and apparently at an alarming rate. The exponential increase continues until eventually, the growth hits some terminating limit which is appropriately referred to in engineering terms as "a hard ceiling" [16]. For a moment in time progress is then suspended. Sometimes the encountered limit is a hard one – literally. Such limits are most often dictated by the constraints of physics and despite all of our best human efforts we are not, at least for the moment, able to push further along such a specific technological path. At other times, the limit is less a physical one than it is a social one. So, for example, as we can see from Figure 10-4, *Titanic's* demise inhibits ship growth but only temporarily. It certainly does not halt it completely. As we know, ships can practically be much larger than the *Titanic*. Physically, both in theory and in practice, ships can be well over an order of magnitude greater. Such growth was indeed actually in evidence even after April 1912 [17].

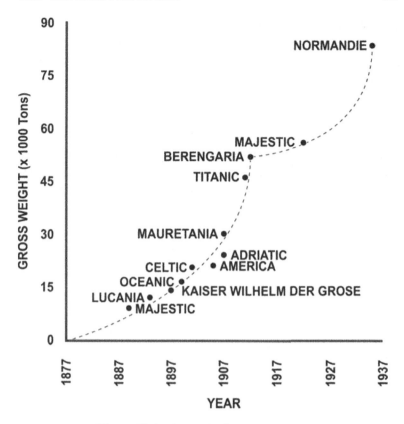

Figure 10-4: *The growth of liner tonnage by year*

Sometimes it is the combination of both the approaching intrinsic natural limit as well as the social constraints that combine to create a momentary stopping point. Often epochs of slowing or cessation have to do with the growth of another superior technology; a process known as supersession. This means that the technology at hand is either overtaken by or on occasion, completely replaced by another "species" of technology. This is very much the case with airship technology, as shown in Figure 10-5. In this particular case, the vivid images of *Hindenburg's* failure and the presence of a heavier-than-air alternative acted to prejudice the paying public away from airships. It was also predicated on the fact that the public perception of risk is almost always skewed away from the rational value [18].Perhaps it was because airships were still very much a 'luxury' transport that the public's face turned away from them and their stately capacities. But we can and should, ask was this rejection not highly similar in many ways to the implicit rejection of the wind-driven "*Fighting Temeraire?*"

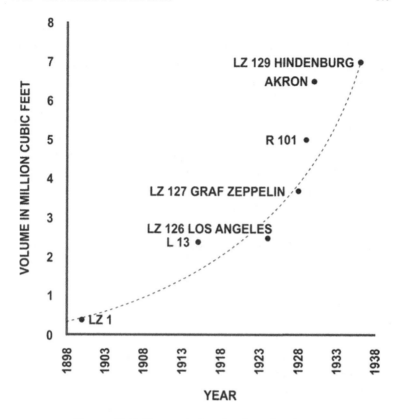

Figure 10-5: *The growth of rigid airship volume by year*

The crucial form of development here is the curve which can be described as an exponential explosion [19]. It characterizes not simply the transports I have presented but most, if not all other facets of technology. In aviation terms, such technologies "*push the envelope*" and we ride each particular wave of progress until we are either thrown by it, the wave naturally plays itself out, or it is in its turn over-ridden by the next wave of technology [20].

These self-same effects are further evident in many alternative examples. Such technical representations can be seen in the United States especially in terms of two vital infrastructure technologies. The first is nuclear power industry which suffered the spectacular but essentially "harmless" accident of *Three-Mile Island* (TMI) [21]. As the subsequent disasters at *Chernobyl* and *Fukushima* have shown, the incident at *Three-Mile Island* could have been worse, much worse. Nuclear power generation is beset by the specter of radiation. This potent but invisible threat is feared by the general public in America to such an extent that TMI produced a grinding halt to nuclear power plant development. It is only now, essentially a generation later, that the United States is starting to re-dip its collective toes in the nuclear pool. Spectacular failure induces a mass misperception of risk and generally politicians and policies are influenced accordingly. A similar pattern was evident in the United States in terms of dam projects. The startling and memorable failure of the *Grand Teton* dam produced a highly similar foreshadowing of the later TMI fallout. To the degree that we are over-loving of our successful technologies, we are also disproportionately wary after any of their failures. However, this is simply a reflection of our intrinsic capacities for trust and mistrust as exhibited by humans individually [22].

10.3. From Stonehenge to Space Station

Throughout this work, I have emphasized the nature, the characteristics, the structures, and the functions of the respective transports I have identified. I now want to claim a degree of commonality across *all* technology. From the Great Pyramid to the Hubble Telescope; stripped of all their physical enclothment they each subserve the same basic function. Each voyage, so undertaken, is contingent upon both the form of transport and on the nature of the individual so transported; but meshed together. In their very essence, these transports are always creations and conceptions of the mind. They serve fundamentally to change the perception and the cognition of the exposed individual. The true origin and destination always reside in the mind of each individual passenger. The hope remains that our technology will take us beyond where we presently are, beyond ourselves even, to recapture a happiness we believe now lost. It is in this way that 'transports of delight' are conveyances for 'moments of wonder.' Moments in time when we become just a little more than the physical animal we actually are and glimpse that which we might become. That destination is perhaps expressed most poignantly in Houseman's evocative stanza [23]:

> *"That is the land of lost content*
> *I see it shining plain*
> *The happy highways where I went*
> *And cannot come again."*

Our respective 'transports' then are not limited only to the individual technologies I have chosen to illustrate. Perhaps, the more interesting question is whether we can conclude that all technologies serve as 'transports' in one fashion or another? If we look at our modern-day technologies such as computer systems and personal communication devices, are not their essential purpose to transpose our consciousness in space and time? As we reach out from our own individual and unique idiographic point of consciousness, are we not using technical orthotics to expand our own presence into the Universe [24]? But are we modern-day humans singular or unusual in this respect? I would argue not. Actually, I would suggest we have been on the self-same path of progress throughout the whole of human existence. It is a vector of progress I epitomize and explore here as our journey from Stonehenge to Space Station.

Figure 10.5: *The ancient wonder that is Stonehenge*

I think that by this time you will be aware of my love of story as well as my pre-occupation with individual historical monuments (technologies) and their associated narratives. Thus as with all of the examples in the present text, I could have chosen from amongst many of the ancient wonders of human construction; from the Pyramids of Giza to Angkor Wat, from

Macchu Pichu to Baalbeck [25]. For convenience and for preference I have selected, a polemical start-point, the southern English Monument of Stonehenge.

10.4. The Quintessential Stone Circle

Stonehenge is thought to date back as far as 3100 B.C but there may have been some presence on this site even back as far as 6,500 years ago. The earliest elements of the monument were a circle and ditch approximately 330 ft in diameter. Stonehenge was built, rather like our gothic cathedrals, in a series of stages perhaps over as much as 1500 years. Early timber fabrications were superseded by the stone constructions we still see today. Somewhere between three and four thousand years ago, the bluestones were purportedly brought from the Preseli Mountains in Wales and erected as a series of uprights inside the horseshoe of sarsens [26]. As might well be imagined, the true history of a monument of this age is shrouded in both mystery and history and, at the same time, it is wrapped in local and national myth. For example, in his reference to the tales of King Arthur, the writer Geoffrey of Monmouth [27] proposed that Stonehenge represented the handiwork of the wizard Merlin. The originators of this creation have not been constrained to just wizards alone. Stonehenge has also been variously attributed to the Egyptians, the Greeks, the Romans, and the Saxons to name but a few. The antiquarian John Aubrey attributed the origin of the monument to the Druids, a postulation that was subsequently embraced by others such as William Stukeley who believed Stonehenge was a temple to the sun. Today, we are a little more conservative in our purported identifications of the original builders but different hypotheses still abound.

If the identity of the builders is a contentious issue, the purpose of the monument has been the subject of almost endless speculation. One of the most popular theories is that promulgated by Gerald Hawkins [28] which is that the circle functions as some form of astronomical calendar. Here, various parts of the circle are apparently aligned with distinguishing events such as equinoxes, solstices, and eclipses. While this celestial almanac idea is a popular notion, it is necessarily the case that any terrestrial circle must line up with certain celestial events, at least in some fashion. It is then only the perceived pattern and observed frequency of these associations that must bear the weight of persuasion. Thus, the astronomical theory may be a sufficient explanation but it cannot rise to the level of an exclusive account. Until we finally understand what was in the minds of the various builders over the years.

The celestial calendar hypothesis is not an exclusive explanation because Stonehenge has also been identified as a site of worship, a burial ground [29] and the termination of certain megalith landscape architectural features spread across the countryside of southern England. Of course, none of these explanations necessarily exclude or preclude any of the others. Therefore, I cannot say with any degree of certainty which of these offered suggestions is "right", or even which one represents the strongest or most likely candidate. However, what I can propose is that hovering behind each and every one of these nominal explanations is the clear and evident function of Stonehenge as a '*Transport of Delight.*'

However, we do not only have to look back into the past to find such examples. This is because we produce these transports in each and every generation. One illustration from our own time is the achievement of the orbiting International Space Station (ISS). One nominal step on the road back to the Moon and on our initial journey to Mars, the ISS serves as a research platform that enhances our knowledge and understanding of space and the limits and constraints of our part in it. It is both a technical wonder and a 'transport of delight' but one that, sadly, few have been able to share on a personal level. [30] Thus, I would make a rather strong, if somewhat nebulous assertion, that all of human technology from earliest tools and machines to the most modern, sophisticated technological each carry this common imprimatur. They transport us toward wonder.

10.5. A Summary of Transportation

In this book, I have presented some thoughts and ideas that have come from pondering the relationship between different 'transports' that I have described and discussed. I want to make one final observation, not simply about the objects that have been the focus here, but about linkages, networks, and reticulations in general. Like all human constructions, the "cathedrals" I have referred to here are basically ideas. They are, like all human constructions, a vision made concrete. I have often asked myself whether the purpose of technology is eventually to provide such instant material gratification to all of the sensory needs, dreams, and desires of all human beings. In essence, to make momentary desires real. I hope not.

Unfortunately, it seems that a profit-driven capitalist society is predicated upon rampant consumerism. The associated function of advertising appears to be the conversion of each and every dilettante desire into a perceived need. Human beings need oxygen, food, and shelter. We do not need electronic cocktail shakers. Contrary to popular opinion, advertising professionals are human also. So it is, in essence, we ourselves who are selling ourselves down this path of every greater consumption and material slavery [31]. Paradoxically, and despite advertising's intrinsic and sometimes explicit claim to the contrary, consumption does not convert to happiness. The false and reprehensible use of the threat of poverty, disease or ill-health to stimulate such consumption is the ultimate crime of species-level self-abuse. Technology should serve to support the needs of all, not the greed of some. Unfortunately, technology has been perverted to stimulate and translate desire to need. The resulting societal angst, frustration, and dissatisfaction is the price we continue to pay for accepting this lie [32]. When we sublimate wander to profit, we engage in the ultimate technological betrayal.

Writing this book has helped me to see that technology can be used to realize true visions. Technology can serve to promote the very best in us beyond our greedy and fearful worst. I have asked in this text that we each pursue our own vision. I cannot see what you see because I do not know what you know. It is only when we move from the role of passive consumer to active producer that the whole vision of humankind can become manifest [33]. Opportunities provided by the web and internet are outlets for such visions as never before. The power of each individual's vision will, as ever, be mediated by the judgments of others. However, the sight of a world where too many voices are 'dumb' and only a vociferous "few" can be heard presents the specter that many become a permanent and pathologic state of our species. As we are biologically free to procreate, surely we must be cognitively free to create.

If technology is truly symbiotic with human evolution, we need the diversity of creation to evolve. Evolution with intent is what I mean by the term 'teleologics.' It is a term I have coined for an intentional 'science of purpose' [34]. I have asserted elsewhere that 'purpose predicates process' and without some longer term vision of what our purpose effectively is, we are adrift indeed. Underspecified but helpful frameworks have been provided by the better facets of revealed religions. However, these tenets are now centuries in abeyance and increasingly fail to embrace and comprehend the full range of technological advances. They are progressively more anachronistic and somewhat like their now sadly devalued "cocoons" (the cathedrals we have been considering) they seem bereft of new and relevant ideals in a world of ever-increasing change. As we become collectively less comfortable in these old and ill-fitting theological clothes, we need to engender a new renaissance of purpose. That purpose has to be founded on a pragmatic recognition of our absolute ignorance of sources of purpose other than our own actions. Into this lacuna we have to inject a degree of collective responsibility and form our own united human purpose and its manifestation in technological advance. It is only in this way that we will evade the extinction of civilization; or as I have termed it incipient -*civicide*.

Reference Notes: Transports of Delight

[1] Cox, M. (2006). *The meaning of night.* (p. 249), W.W. Norton: New York.

[2] Such acceleration is epitomized in the well-known expression of Moore's Law, see: Moore, G.E. (1965). Cramming more components onto integrated circuits. *Electronics*, 114–117, April 19 which speculates upon the doubling rate of computational capacity at a fixed rate initially estimated to be eighteen months in duration.

[3] Hancock, P.A. (2005). Time and the privileged observer. *Kronoscope*, 5 (2), 176-191.

[4] See e.g.: Von Simson, O.G. (1988). *The Gothic Cathedral.* Princeton University Press: Princeton, NJ.

[5] Interestingly, one can make the self-same argument for personal automobiles being an extension of the uterus. Our own personal cars are a source of "freedom" in which they take us from origin to destination, climate-isolated from the outside and a warm, nurturing space inside. I have never really seen them as symbols of masculinity but much more of femininity. Indeed, if they are a surrogate representation of our mother, no wonder so many individuals get angry and suffer 'road rage' when their mother substitute is threatened! (and see: *My Mother the Car.* http://en.wikipedia.org/wiki/My_Mother_the_Car

[6] see: Skelton, R.A., Marston, T.E., & Painter, G.D. (1965). with Introduction by Alexander O. Vietor. The Vinland Map and the Tartar Relation Yale University Press, 1965 (VMTR). Reprinted in 1995 with new prefatory essays by Painter, Wilcomb E. Washburn, Thomas A Cahill and Bruce H. Kusko, and Laurence C. Witten II (VMTR95), but original pagination retained in body. See also: Larner, J. (1999). *Marco Polo and the discovery of the world*. Yale University Press: New Haven, CT.

[7] Kujawa-Holbrook, S.A. (2013). *Pilgrimage: The sacred art journey to the center of the heart*. Jewish Light: Woodstock, VT.

[8] Of course, this also pertained to emigrants to other places across the world; and see: Plowman, P. (1994). *Emigrant ship to luxury liner*. University of New South Wales Press: Sydney, Australia.

[9] Bailyn, B. (2003). *To begin the world anew*. Vintage: New York.

[10] see: http://en.wikipedia.org/wiki/Hindenburg-class_airship.

[11] And see: Gleick, J. (2000). *Faster: The acceleration of just about everything*. Vintage: New York.

[12] For example, Margret Mather talked of: *"a lift and pull upward, an indescribable feeling of lightness and buoyancy,"* and frequent traveler Leonhard Adelt was further impressed by *Hindenburg's* capacity to respond to storms, to him, *Hindenburg* "*glided as smoothly through the black storm clouds as though it were a calm, moonlit night.*" Nelson Morris noted that: "*This is the most interesting experience I have ever had in my life. . . . I cannot find words to properly express the sensation.* "Perhaps it is true that, as Alexander Chase has said:: "*Lovers of air travel find it exhilarating to hang poised between the illusion of immortality and the fact of death.*"

[13] Quotation from the airship company's own pamphlet: *"Airship Voyages Made Easy."*

[14] see the interesting text by: Halberstam, J. (2011). *The queer art of failure*. Duke University Press: Durham, NC.

[15] This is also evident in contemporary growth of, for example, computing power. See Moore, G. (1965). *Cramming more components onto integrated circuits*. New York: McGraw-Hill.

[16] See discussions in articles such as: Whiting, E., Ochsendorf , J., & Durand, F. (2009). Procedural modeling of structurally-sound masonry buildings. *ACM SIGGRAPH Conference Proceedings Trans. Graphics, 28* (5), 112.

[17] The RMS Berengaria was laid down before the Titanic disaster and was of a greater length (879ft.) and tonnage (52,226) than Titanic. She was the next along the line of progress but after her we see the inflexion in the curve of development.
See: *www.norwayheritage.com/p_ship.asp?sh=beren*.

[18] See: Kahneman, D. (2011). *Thinking, fast and slow*. Farrar, Strauss, & Giroux: New York. See also: Mahaffey, J. (2014). *Atomic accidents: A history of nuclear meltdowns and disasters: From the Ozark Mountains to Fukushima*. Pegasus: New York.

[19] Kurzweil, R. (2005). *The singularity is near: When humans transcend biology*. Penguin: New York.

[20] Mager, N.H. (1987). *The Kondratieff waves*. Praeger: New York.

[21] See: Walker, S.J. (2004). *Three Mile Island: A nuclear crisis in historical perspective*. Berkeley: University of California Press. See also: Perrow, C.A. (1984). *Normal accidents: Living with high rick technologies*. Basic Books: New York.

[22] See: Hancock, P.A., Billings, D.R., & Oleson, K.E. (2011). Can you trust your robot? *Ergonomics in Design, 19* (3), 24-29. And see also: Hancock, P.A., Billings, D.R., Olsen, K., Chen, J.Y.C., de Visser, E.J., & Parasuraman, R. (2011). A meta-analysis of factors impacting trust in human-robot interaction. *Human Factors, 53* (5), 517-527.

[23] Housman, A.E. (1919). http://www.gutenberg.org/files/5720/5720-h/5720-h.htm.

[24] Hancock, P.A. (2009). *Mind, machine and morality*. Ashgate: Chichester, England.

[25] Mannikka, E. (2000). *Angkor Wat: Time, space, and kingship*. University of Hawaii Press: Honolulu. Alouf, M.M. (1999). *History of Baalbek*. Book Tree: San Diego. Reinhard, J. (2007). *Machu Picchu: Exploring an ancient sacred center*. The Cotsen Institute of Archeology Press: Los Angeles.

[26] Langdon, R.J. (2013) *The Stonehenge enigma*. ABC Publishing: Brighton.

[27] Geoffrey of Monmouth (1966). *The history of the kings of Britain.* Penguin: London. (original Publication in 1136 AD).

[28] Hawkins, G.S. (1965). *Stonehenge decoded.* Doubleday Books; Garden City, Newyork.

[29] Thomas, J. (2008). Dates for Stonehenge burials signify long use as cemetery. Retrieved from: http://witcombe.sbc.edu/sacredplaces/stonehenge.html.

[30] See: Hancock P.A., & Drury, C.G. (2011).   Does Human Factors/Ergonomics contribute to the quality of life? *Theoretical Issues in Ergonomic Science, 12,* 1-11. And: Dekker, S.W.A., Hancock, P.A., & Wilkin, P. (2013). Ergonomics and the humanities: Ethically engineering sustainable systems. *Ergonomics, 56* (3), 357-364.

[31] I should note that now, regular visitors are not allowed to actually enter the central circles of Stonehenge but must circle around the outside. Again, this is symbolic of the removal of "normal" people from the center of action. Of course, this like all similar actuons is done for the most laudable of reasons – in this case preservation. And see: Hardy, T. (1891) *Tess of the d'urbervilles.* Osgood, McIlvaine: London.

[32] Hancock, P.A. (2012). Notre trahison des clercs: Implicit aspiration, explicit exploitation. In: R.W. Proctor, and E.J. Capaldi, (Eds.), *Psychology of Science: Implicit and Explicit Reasoning.* (pp. 479-495), New York: Oxford University Press.

[33] Illich, I. (1973). *Tools for conviviality.* Boyars: London.

[34] Hancock, P.A. (1996). Teleology for technology. In: R. Parasuraman and M. Mouloua. (Eds). *Automation and human performance: Theory and applications.* (pp. 461-497), Erlbaum, Hillsdale, New Jersey.

❖ ❖ ❖

"The purpose of life is a life of purpose." [1]

11. AUTOBIOMIMESIS

ONE JOURNEY TO ULTIMATE PURPOSE

11.1. Mimetic Principles

Now that we have come to our final chapter, I will look exclusively at the issue of purpose in technology. As we have already seen, one primary purpose for, and indeed perhaps the origin of technology, is to provide prosthetic replacement for, and/or orthotic extensions to, resident human capabilities [2]. Here, I use the term *prosthetic* primarily to mean the replacement of some pre-existing function. Then, somewhat in contrast, I use the term *orthotic* as representing a technology which underwrites any form of augmentations to such functions. Both *prosthetics* and *orthotics* serve, in their own individual way, to improve upon what humans can achieve alone.

While creating tools by using tools to do so and employing the subsequent elaborates of technology has characterized human beings, it is the case that sometimes, what one individual can achieve alone, or with only very rudimentary tools, is so amazing that other human beings can only conceive of that product as being "magic" [3]. This propensity should remind us of what is probably Arthur C. Clarke's most famous quotation; being as his *"Third Law"* which states: *"Any sufficiently advanced technology is indistinguishable from magic"* [4]. The question I want to resolve here is not simply *how* we perform such technological magic [5] but *why* we look to achieve these magical goals in the first place.

© Springer International Publishing AG 2017
P. Hancock, *Transports of Delight*, DOI 10.1007/978-3-319-55248-4_11

After all, with respect to most living systems, the question of *why*, as apposed to *how* is relatively rarely addressed. As exemplified by the wonderful cartoons of Gary Larson, we even consider it comic when we are shown other members of the living world exhibiting our own unique type of curiosity. Formally then, I am moving here from the notion of *efficient* cause (i.e., the 'how' of existence) [6] to the notion of *final* purpose (i.e., the 'why' of existence) [7].

From the magnification premise of technological influences, as instantiated in prosthetics and orthotics we can see that power plants (nuclear or otherwise) for example, are fundamentally extensions to the bodily metabolism. Both of these mechanisms, one (metabolism) being nominally 'natural' the other, the power-station operations nominally 'artificial', each act to produce useful energy. Likewise, levers expand the power of muscular action by magnifying the effect of the initial energy which is invested; this is the case whether such energy was originally generated by the person themself or by any proximal form of technology they are using [8]. As additional examples we might cite the fact that telescopes and microscopes magnify the abilities of the unaided eye to perceive more efficiently and effectively in the visible spectrum. As orthotics, infra-red and ultra-violet scanning technologies enable visualization of frequencies beyond the visible-light spectrum. These mechanical replacements and elaborations are based on what are termed *mimetic* principles. Mimetics guide technical developments contingent upon existing natural capacities. In principle all mimetic principles recapitulate nature, wherein advantage is taken of pre-existing architectures. In living systems, such principles represent a sub-category of the larger principle of mimesis, which is labeled '*biomimesis.*' I should note that most of the aspects of mimesis, which concern me here, are actually *biomimetic* in nature.

It is important to observe that nature often exploits or copies herself. That is, nature often replicates and benefits from her own past successes. Nature is a natural self-plagiarist; a role that is, somewhat paradoxically, now frowned upon in current scientific circles [9]. For having once invented a useful strategy that supports the sustenance and spreading of life, nature then looks to incorporate these self-same solutions into other evolving life forms. The same thing happens of course with human technologies. Having once invented something useful, like the wheel, we incorporate not only wheels themselves but the very idea of a *wheel* itself into much more complicated designs. Within the sub-category of biomimesis there lies yet a further embedded set of principles concerning human action with respect to the technology that we ourselves create. I have termed this latter process "*auto- biomimesis.*"

Auto-biomimetics are uniquely human activities, which are represented by the design of, the creation of, the use of, and the aspirations for technology. Autobiomimetic principles derive form the underlining premises that humans are the only species to use tools to makes tools. *Magnification* is one central organizing drive embedded into the human purposes of technology and this is therefore basically an *autobimimetic* force which is extended in the search for ultimate purpose. *Replication* and *expansion* are clearly (orthotic) drivers of evolutionary vector. However, the physical sustenance and prolongation of life cannot, and should not, alone be our sole motivation. While autobiomimetic strategies certainly underwrite physical requirements such as food and shelter [10].

and while their various elaborations also serve to support our capacity to generate a highly inter-dependent civilization, this is also not all that technology either can or should do. Before we attack this final question as to what ultimate purpose can be, we need to evaluate the caution that pathologies which derive from such magnified technical capacities present. These are essentially technological *"stations of the cross"* that act as warnings of maladaptive paths of progress. This survey then acts as a warning as to what can result when disruptive capacities of humans are replaced, replicated, magnified, and expanded under the auspices of technological power. As a result of the enormous magnifying effect of all of technology, such   dysfunctions now threaten our very existence as a species [11].

11.2 . Pathologies in Technology

Autobiomimesis then need not necessarily be progressive nor indeed constructive, even when it does succeed. As one form of technology, a gun is as surely an extension of a fist as other interpretive magnifications of any human capacity that I have considered here. However, even when such a technological tool "works" there is no necessarily good or morally positive outcome from its use [12]. Weather this statement could be true of all technology is a proposition we have to consider. All current 'extensive' technologies replicate parts of the human body but they are, in their most fundamental essence, elaborations of the activity of the human brain. So, we can view all facets of technology essentially as extensions to sensation, perception, cognition, decision-making, and *then* subsequent effector actions.

Technology is the natural extension of past and present brains; their ideas turned inside out.

In a reflexive way, we can also see how technology itself has always been used as a metaphor to describe and 'explain' the workings of the brain itself. For example, each sequentially dominant metaphor for how the brain operates has always used the respective system that was the then highest existing form of contemporary technology. Early brain metaphors evolved from original propositions that the brain was an advanced plumbing system. The metaphor then evolved through the idea of the brain as a form of superior telegraph exchange just when that technology predominated in the middle of the nineteenth century. This line of metaphorical comparison has culminated in our own times with the modern pre-occupation with the computer metaphor for both mind and brain (as software and hardware respectively) [13]. It was the savant Richard Gregory who intimated that this comparison might necessarily persist because the human brain is the most complex operating system that we are currently aware of in the universe. Hence, it necessarily has to be compared to the most complex technical capacity that we possess; essentially the most apposite model we possess in each successive era [14].

We know however, that in the same way metabolism can fail and muscles can atrophy, there are a rather large number of ways in which the brain itself can malfunction. What is interesting here is that we can see manifestations of many of these brain illnesses and syndromes when we extend and magnify them outward into the world where they become technological failures. For example, Parkinson's disease is a disability of the motor system and is predominately characterized by symptomatology such as an uncontrolled tremor of one or more of the limbs. This is a parallel in human-machine systems with respect to a problem that is labelled 'pilot-induced oscillation.' In this technological, recapitulation of Parkinsonism, there is an inherent and maladaptive time-lag involved in the operations of a complex system. As the name suggests, this is most often epitomized in the operations of high-performance single-seat jet fighter aircraft; or even more notoriously, early space shuttle operations. Pilot-induced oscillation occurs when the pilot provides a control-stick input to the aircraft and, like our own personal automobile, they expect to see an immediate effect of that input (e.g., turn [bank] left, right, climb, descend, etc.).

This immediate form of response represents, something that is called a zero time-lag system (not necessarily to be confused with something called zero-order control [15]). However, if this initial pilot input is delayed by a sufficient interval of time (and here we are talking in the order of only a few tenths of a second), then the pilot not feeling the effect of their first input, starts to input a second, corrective command, just as the first one takes effect. The pilot then reacts to this alteration of the aircraft's attitude with the entry of yet another following compensatory action. These sequential responses now begin to cascade one upon the other. The result is a series of oscillations or tremors as the overall system fights to retain stability. In reality this is more of a design-induced oscillation than it is a pilot-induced problem. This is because the original design provides a system response that is incommensurate with human reaction and response time, i.e., they are desynchronized. A graphic and visual example of this can be seen happening at the initial landing of the space shuttle [16].

Here, the pilots Young and Crippen, are fighting to land what is an inherently unstable system. That they resolved this is a testament to the adaptive capacity of skillful humans and an exhibition of these pilots' own personal abilities. So failures in the brain, and of course Parkinson's is much more complex than a pure delay can be seen as subsequently being replicated and magnified through poor technological systems design. Thus, technology is by no means a ubiquitous "success".

It is a categorical mistake however, to see this one particular form of human–machine pathology as only occurring in one specific type of system or domain such as advanced aircraft. Indeed, when teams of people act together in a group, in which inputs are discordantly slowed or accelerated, by one person or another they can strike the self-same form of instability. These disruptions happen in applied realms as diverse as management systems, health-care operations, banking, stock market trading, as well as many, many others. These uncontrolled or un-damped oscillations are nearly always problematic when they occur [17]. But variations in timing as extended by technology are by no means the whole story of maladaptive brain magnifications.

11.3. Further Eversions of the Brain

A second form of brain mal-adaption which is expressed when we employ autobiomimetics expansion is schizophrenia. In schizophrenia, one single individual experiences a fractured sense of self, expressed as a disunity of consciousness. This manifests itself in various forms in human-machine operations and is sometimes referred to as mode error awareness; being a sub-category of a larger issue known as situation awareness [18]. As with any individual's schizophrenia, mode error awareness can lead to very problematic outcomes. Continuing with the earlier mentioned aviation theme, we can see the catastrophic results which occur when pilots who are operating a large modern-day fly-by-wire aircraft 'lose' their awareness of what needs to be known and what needs to be done for safe and effective operations [19]. As the number of possible system (aircraft) states increase, then the propensity to become "lost" in these various states also increases. Mode errors have been implicated in, and identified as the cause of, many aircraft crashes. As modern aircraft become ever-more complicated and can rightly be considered flying super-computers, the probability of experiencing mode awareness error also increases, and thus has to be guarded against.

Along with schizophrenia, we can also see paranoia expressed in extensions of human-machine technologies. The critical element affected in this maladaptation is *trust*. Appropriate system operations require the appropriate levels of trust [20]. However, many problems arise when trust is miscalibrated. Over trust or over-reliance on automated technologies leads to *"automation complacency."* In contrast, under-trust leads to system disuse and *"automation neglect"*. This particular maladaptation can be especially problematic because of the vast and increasing rate of growth of the automatic nature of many of our modern-day technologies [21]. And when we consciously doubt the relatability of our technologies we begin down a grim road indeed. Although we must acknowledge that some of these technologies should indeed be mistrusted. The final example, from the many I could have looked at, that I have chosen to illustrate here is attention-deficit disorder (ADD). I could, of course, have run the whole gamut of the DSM-5 [22] and identified each of the respective problems and issues; explaining how each in their turn are transmuted into technical mal-adaptations. However, I will content myself here with this one last case.

With ADD, the individual is thought to be unable to properly fixate their attention on appropriate sources of information for the required period of time. However, as humans become more and more guardians of semi-automated systems and eventually monitors of fully automated ones, the issue of prolonged attention has become ever-more important. In human-technology research this capacity is known as *vigilance* or *sustained attention* [23]. Failure of this capacity is frequently implicated in the etiology of large-scale technical disasters [24]. The main point that I want to emphasize is that technological extensions to human cognition do not always create either 'wonder' or success. Far from it. In many circumstances, these technical magnifications create both physical and mental 'nightmares.' When we evert the brain and empty it into the willing receptacle of the environment, it sculpts all the possible products of cognition, however 'good' or 'bad' these visions might be [25]. Thus, *'wonder'* here is a relative term. Finally, in this respect, we must also consider the proposition that the collective actions of our own species are not necessarily "good" anyway. In fact, it could be argued that human beings, in degrading the collective living environment, are actually agents of destruction. Such thoughts threaten the standard narrative in which we place ourselves in the role of the shining hero of the on-going story [26].

Technology moves our world. It is the well-spring of our social civilization. However, it is at one and the same time the source of our greatest threat also. We all experience the effects of technologies every day. Not only do technologies augment each one of us, they are our primary human ecosystem. However, anyone who has battled a telephone menu-system, been put "*on-hold*" by a computer, have been "*recorded for quality purposes*," had a total systems crash, or spent five or more unproductive minutes pounding on an unresponsive keyboard will be aware that such technologies still remain far from perfect [27]. While most technologies actually work surprisingly well most of the time, when they fail the results are not necessarily just frustrating; failures are often fatal. Fatal now not just for one or two proximal individuals who happen to be in the "wrong place at the wrong time." These failures now threaten the well-being and existence of whole groups, cities, nations, and most recently our whole species.

11.4. It's All in Your Hands

I hope that I have been able to demonstrate here that the landscape around us is littered with technologies, each waxing and waning in their own respective niche in space and time. As you now are personally unlocking the hands of time, I cannot pass over the fact that the very book you are holding in those hands is itself both a dying technology as well as a quintessential '*transport of delight*' [28]. There may be some who are reading this work via some electronic medium such as a "*Kindle*" or "*Nook*" [29] and perhaps they have never gotten "lost' in a traditional book. This would be a tragedy because like video-games, theme parks and other types of entertainment, getting 'lost' is merely code for, and one of the principle hallmarks of, the wonder of transportation. 'Lost' in these terms means not here and not now but somewhere else and sometime else; and such relocations are mostly beneficial [30]. The traditional book is, sadly, now slowly disappearing. As a technology it requires a material medium which, for a single physical book, is largely fixed and unchanging. It also takes considerable physical storage space as each of the major libraries of the world demonstrate. For myself, as a bibliophile who both enjoys and collects books, these are vulnerable friends and valued companions with whom I visit and reminisce on numerous occasions. For purely profit-driven publishers, they are a progressively inefficient way of generating revenue.

An electronic window that can convey access to the same information as a book is a more efficient income producing conduit. And, like virtually everything else in our present lives, we are being forced to bow down to the ever-more meaningless pursuit of profit. Thus, rather than seeking and pursing true human-centered and human-elevating technologies [31], what viral capitalism requires and embraces are endless streams of such gizmos and gadgets that produce the greatest profit-margin. It is the leitmotif tragedy of our times that this pattern is imposed on 99.99% of the population while a vanishingly small percentage of the others claim a bespoke, but often equally unfulfilled life [32]. It is a recipe for planet-wide technical and social disaster; an eventuality that I have termed *civicide* [33]. I hope that you are reading my work in traditional book form and I hope that, if only for a little while, you have been 'lost' in it. For, as I said at the very beginning, wonder must be self-generated to be a truly life-altering experience. In the most literal sense, it is now in your hands.

11.5.   The Celestial City

For any transport system its fundamental function must surely be to help accomplish a desired journey. Where there is a journey it is necessarily defined by an origin, a transition and a destination. So what, if any, is our human destination? For early Christian pilgrims this was the glimpse of Jerusalem, the true eternal city. For those limited to their local church or place of worship like a cathedral, it was a brief chance to glimpse the Almighty; the epitome of the eternal. For those in steerage on a trans-oceanic liner, or those even undergoing the enforced emigration of criminal transportation it was the first sight of their new home. Like all destinations these are surely places of hope and of promise. As with the redeemed Christian in Bunyan's *"Pilgrim's Progress"* [34] this represents the prospect and achievement of a new life; in essence re-birth. For those used to opulence and luxury, it was as much about the passage itself as it was the reaching a new place. Sadly, extreme wealth can be almost as debilitating to human beings as extreme poverty—but not quite. Perhaps our eventual human goal will be represented by our seamless merging with our own self-created technology; the full realization of what I have termed *self-symbiosis* [35].

In all of these cases, it is the presence of hope that is the vital common element. And rather like buying a modern-day lottery ticket that we know that statistically must lose, what we really seek is the transient feeling of what could be. This is the key to understanding that the true destination of technology and human beings is hope itself—an experience of wonder [36]. Unfortunately, reality, even when we achieve our dreams, often does not match either our hope or our aspiration [37]. Better to journey on in hope than to arrive in despair.

Reference Notes: Autobiomimesis and Ultimate Purpose

[1] Original quote from Robert Bryne. Consider also this *"The mystery of human existence lies not in just staying alive, but in finding something to live for."* Dostoyevsky, F. (1880). *The brothers Karamazov.* (1952 Edition), Spartan Press: Raleigh, NC.

[2] Prosthetics can be defined as: *"a device, either external or implanted, that substitutes for or supplements a missing or defective part of the body."* In contrast, orthotics have been defined as: *"The science that deals with the use of specialized mechanical devices to support or supplement weakened or abnormal joints or limbs."*

[3] Hancock, P.A. (2012). The making of myth: Edward Leedskalnin and the Coral Castle. *Skeptic, 18* (1), 44-50.

[4] Clarke, A.C. (1962). *Hazards of prophecy: The failure of imagination. In: Profiles of the future: an inquiry into the limits of the possible.* (14-36). New York: Harper & Row.

[5] It is this level of magic, or *magicke* that we now term mechanical engineering. In the sixteenth century the necromancer Dr. John Dee termed it thaumaturgike. And see: Dee, J. (1570) *Mathematical preface.* London: Kessinger Publishing, LLC

[6] And see: Kauffman, S. (1993). *The origins of order.* Oxford: Oxford University Press.

[7] Aristotle's Four Causes. More information can be found at: http://en.wikipedia.org/wiki/Four_causes.

[8] John Henry vs. the Steam Hammer.

[9] This is limited to expressed word patterns that should not be repeated, not ideas per se-as yet.

[10] Rothenberg, D. (1993). *Hand's end: Technology and the limits of nature.* University of California Press: Berkley.

[11] See, e.g., Maslow, A.H, & Herzeberg, A. (1954). Hierarchy of needs. In: A.H Maslow. (Ed.). *Motivation and personality.* Harper, New York.

[12] Hancock, P.A. (2014). *Solutions to iatrogenic civicide: Changing minds and machine to change the future.* Unpublished Manuscript. See also the incredible personal price paid by Robert Oppenheimer as he realized that he has invented the ultimate destructive extension to the human clenched fist; http://www.youtube.com/watch?v=lb13ynu3Iac. We also now smile rather knowingly at Alfred Nobel's hopeful but sadly naïve view of his own extension to the human fist where he says; *"Perhaps my factories will put an end to war sooner than your congresses: on the day that two army corps can mutually annihilate each other in a second, all civilized nations will surely recoil with horror and disband their troops."* http://www.nobelprize.org/alfred_nobel/biographical/articles/tagil/, Hancock, P.A. (2009). *Mind, machine and morality.* Chichester: Ashgate. c.f., Barreett, P.M. (2012). *Glock: The rise of the America's gun.* Broadway: New York.

[13] Gregory, R.L. (1981). *Mind in science: A history of explanations in psychology and physics.* New York: Cambridge University Press.

[14] See also: Russell, B. (1912). *The problems of philosophy.* Home University Lansbury (First published Oxford University Press, 1959). This notion was discussed by myself and Gregory on one of his visits to

America. He challenged me to find any more adequate metaphor to which I replied that the universe itself might perhaps serve in this capacity. The argument which then followed was both lively and informative. Simply, Gregory is missed. For intriguing work on modeling and its fundamental limits see: Sheridan, T. (2014). *What is God? Can Religion be modeled.* New Academic Publishing: Washington, D.C.

[15] Jagacinski, R.J. & Flach, J.M. (2003). *Control theory for humans: Quantitative approaches to modeling performance.* Mahwah, New Jersey: Erlbaum.

[16] http://www.youtube.com/watch?v=O8fnUY2IFAw In the case of the Space Shuttle, the circumstances were a little more complicated than a pure lag in the system.

[17] Gottlieb, I.M. (1987). *Understanding oscillators.* Tab Books: Blue Ridge, PA.

[18] See: Smith, K., & Hancock, P.A. (1995). Situation awareness is adaptive, externally-directed consciousness. *Human Factors, 37* (1), 137-148.

[19] Sarter, N. and Woods, D.D. (1995). How in the world did we ever get into that mode? Mode error and awareness in supervisory control. *Human Factors, 37* (1), 5-19.

[20] Lee, J.D., & See, K.A. (2004). Trust in automation: Designing for appropriate reliance. *Human Factors, 46* (1), 50-80.

[21] See: Hancock, P.A., Billings, D.R., & Oleson, K.E. (2011). Can you trust your robot? *Ergonomics in Design, 19* (3), 24-29. Also: Parasuraman, R., & Riley, V. (1997). Humans and automation: Use, misuse, disuse, abuse. *Human Factors, 39* (2), 230-253. As well as: Hancock, P.A., Billings, D.R., Olsen, K., Chen, J.Y.C., de Visser, E.J., & Parasuraman, R.

(2011). A meta-analysis of factors impacting trust in human-robot interaction. *Human Factors, 53* (5), 517-527.

[22] *The Diagnostic and Statistical Manual of Mental Disorders,* Fifth Edition, abbreviated as DSM-5, is the 2013 update to the American Psychiatric Association's (APA) classification and diagnostic tool. And see: http://en.wikipedia.org/wiki/DSM-5.

[23] See: Hancock, P.A. (2013). In search of vigilance: The problem of iatrogenically created psychological phenomena. *American Psychologist, 68* (2), 97-109.

[24] Vigilance failures in large-scale failures. See Casey, S. (2006). *The atomic chef: And other true tales of design, technology, and human error.* Santa Barbara, Calif.: Aegean Pub. & Casey, S. (1998). *Set phasers on stun: And other true tales of design, technology, and human error* (2nd ed.). Santa Barbara: Aegean.

[25] See also Moray N. (1993). Technosophy and humane factors: A personal view. *Ergonomics in Design,* October, 33–39.

[26] See also: Campbell, J. (1972). *The hero with a thousand faces.* Princeton, N.J.: Princeton University Press.

[27] Vicente, K. (2004) *The human factor: Revolutionizing the way people live with technology.* New York: Routledge.

[28] I am, of course, referring to books in general. As to whether my particular book has succeeded in its goal is very much up to you the reader.

[29] These were two differing types of e-reading technologies which, hopefully, will be defunct before the present text also becomes defunct.

[30] Of course, it is not always beneficial to be 'lost.' For example, texting or using a cell-phone while driving is very much an example of when not to be lost in this fashion; and see: Hancock, P.A., Mouloua, M., & Senders, J.W. (2008). On the philosophical foundations of driving distraction and the distracted driver. In: Regan, M.A., Lee, J.D., and Young, K.L. (Eds.). *Driver distraction: Theory, effects and mitigation.* (pp 11-30), CRC Press. Boca Raton, FL, and: Hancock, P.A. (2013). Driven to distraction and back again. In: M.A. Regan, T. Victor, and J. Lee, (Eds.). *Driver distraction and inattention: Advances in research and countermeasures.* (pp. 9-25), Ashgate, Chichester, England.

[31] Illich, I. (1973). *Tools for conviviality.* Harper & Row: New York.

[32] This respective imbalance between resources and the human population was first described by the political scientist Gini and is recognized by an eponymous "Gini Curve," and see: Gini, C. (1912). Variabilita and multibilita. *Studi economico-Giuridici dell'Univ. di Calgliari. 3* (2) 1-158.

[33] Hancock, P.A. (2014). *Solutions to iatrogenic civicide: Changing minds and changing machines to change our future.* Manuscript in Progress.

[34] Bunyan, J. (1678). *The pilgrim's progress from this world to that which is to come; Delivered under the similitude of a dream.* Ponder: Cornhill, London.

[35] Hancock, P.A. (2009). *Mind, machine and morality.* Ashgate: Chichester, England. See Mark Twain also when he comments: *"the human being is a machine. An automatic machine. It is composed of thousands of complex and delicate mechanisms which perform their functions harmoniously and perfectly, in accordance with laws devised for their governance, and over which the man himself has no authority no mastership, no control. For each one of these thousands of mechanisms the Creator has planned on enemy, whose office is to harass it, pester it, persecute it, damage it, afflict it with pairs, and miseries ultimate destruction."*

[36] Campbell, J. (1988). *The power of myth.* Achor: New York. See Also: A Quinas, T. (c1268). De ente et essential (concerning being and essence). Appleton-Century Crofts Inc.; Edition, 1937 G.C. Leckie (Trans.).

[37] Consider also Mark Twain's wonderful pastiche on Heaven, which is surely the goal of many human spirits. Surely, if we believed absolutely in such an eventuality we would fully prepare for it. Specifically concerning the need for harping skills Twain notes that: *"Meantime, every person is playing on a harp -- those millions and millions! -- whereas not more than twenty in the thousands of them could play an instrument in the earth, or ever wanted to. Consider the deafening hurricane of sound -- millions and millions of voices screaming at once and millions and millions of harps gritting their teeth at the same time! I ask you: is it hideous, is it odious, is it horrible?"* Twain, M. (1909). *Letters from Earth.* Harper Perennial (1962 Edition: New York. On a more personal scale, how many times have I passed through Orlando airport witnessing the detritus of what was to be many family's "trip of a lifetime?"

REFERENCES

CHAPTER 1: INTRODUCTION

Burke, J. (1978). *Connections*. New York: Little Brown & Co.

Campbell, J. (1988). *The Power of Myth*. New York: Anchor.

Cox, M. (2006). *The meaning of night*. New York: W.W. Norton.

Gregory, R. (1981). *The mind in science*. London: Penguin.

Knight, B.(2002). *Knight: My story*. New York: Thomas Dunne Books.

Pears, I. (2002). *The dream of Scipio* (p. 263), New York: Riverhead Books.

© Springer International Publishing AG 2017
P. Hancock, *Transports of Delight*, DOI 10.1007/978-3-319-55248-4

CHAPTER 2: GHOSTS OF THE TEMERAIRE

Adkin, M. (2007). *The Trafalgar companion: A guide to history's most famous sea battle and the life of Admiral Lord Nelson*. London: Aurum Press.

Adkins, R. (2005). *Trafalgar: The biography of a battle*. London: Abacus.

Adkins, R. (2006). *Nelson's Trafalgar: the battle that changed the world*. London: Penguin.

Bennett, G. (1977). *The Battle of Trafalgar*. London: Batsford.

Bockenmuhl, M. (2010). *Turner*. Hong Kong: Taschen.

Clayton, T., & Craig, P. (2005). *Trafalgar: The men, the battle, the storm*. London: Hodder & Stoughton.

Corbett, J. S. (1910). *The campaign of Trafalgar*. London: Longmans, Green and Company.

Czisnik, M. (2004). Admiral Nelson's tactics at the Battle of Trafalgar. *History, 89* (296), 549-559.

Egerton, J. (1995). *Turner: The fighting Temeraire*. London: National Gallery.

Fraser, E. (1906). *The enemy at Trafalgar: an account of the battle from eye-witnesses' narratives and letters and despatches from the French and Spanish fleets*. London: Dutton.

Fremont-Barnes, G. (2005). *Trafalgar 1805: Nelson's Crowning Victory* (Vol. 157). Oxford: Osprey Publishing.

Gardiner, R. (2005). *The campaign of Trafalgar 1803-1805*. London: Mercury.

Goodwin, P. (2005). *The Victory*. Norwich: Jarrold Publishing.

Goodwin, P. (2005). *The ships of Trafalgar: The British, French, and Spanish fleets October 1805*. London: Conway Maritime.

Herrmann, L (1975). *Turner: paintings, watercolours, prints and drawing*. London: Phaidon.

Howarth, D. (1969). *Trafalgar: The Nelson touch* (p. 98). Collins: Hammersmith.

Humphries, P. (1995). *On the trail of Turner*. Cadw: Cardiff.

Nicolson, A. (2005). *Men of honour: Trafalgar and the making of the English hero*. London: Harper Collins.

Pocock, T. (2005). *Trafalgar: An eyewitness history*. London: Penguin Classics.

Morris, J. (1968). *Pax Britannica*: The climax of Empire. San Diego: Harvest.

Schom, A. (1990). *Trafalgar: Countdown to Battle, 1803-1805*. New York: Atheneum.

Taylor, A. H. (1950). The battle of Trafalgar. *The Mariner's Mirror, 36* (4), 281-321.

Willis, S. (2010). *The fighting Temeraire: Legend of Trafalgar*. London: Quercus.

Wilson, A. (1980). *Turner and the sublime*. London: British Museum.

CHAPTER 3: WHAT A SIGHT IT IS

Adelt, L. (1937). The last trip of the Hindenburg. *Reader's Digest*, November.

Archbold, R., & Marschall, K. (1994). *Hindenburg: An Illustrated History*. New York: Warner Books.

Becker, B. (1967). *Dreams and realities of the conquests of the skies*. New York: Atheneum.

Botting, D. (1981). *The giant airships*. New York: Time-Life Books.

Brandt, R. (1936). *Mit Luftschiff Hindenburg uber den Atlantik*. Berlin: Scherl.

Brooks, P.W. (1992). *Zeppelin: Rigid airships 1893-1940*. London: Putnam Aeronautical.

Clarke, B. (1961). *The history of Airships*. London: Herbert Jenkins.

Countryman, B. (1982). *R-100 in Canada*. Ontario: Boston Mills.

Dallison, K. (1969). *When Zepplins flew*. New York: Time-Life Books.

Deighton, L., & Schwartzman, A. (1978). *Airshipwreck*. London: Book Club Associates.

Dick, H.G., & Robinson, D.H. (1985). *The golden age of the great passenger airships: Graf Zeppelin & Hindenburg*. Washington, DC: Smithsonian Institution Press.

Eckener, H. (1958). *My Zeppelins*. London: Putnam.

Hancock, P.A. (2001). Five times the Hindenburg! (letter). *Skeptical Inquirer, 25 (4)*, 76-77.

Higham, R. (1961). *The British rigid airship 1908-1931*. Fowler & Co.: London.

Hoehling, A.A. (1962). *Who destroyed the Hindenburg?* Boston: Little, Brown and Company.

Kirschner, E.J. (1957). *The Zeppelin in the atomic age.* Urbana, IL: University of Illinois Press.

Knight, R.W. (1938). *The Hindenburg accident: A comparative digest of the investigations and findings, with the American and translated German reports included.* Bureau of Air Commerce, Safety and Planning Division, United States Department of Commerce. Washington, D.C.

Lace, W.W. (2008). *The Hindenburg disaster of 1937.* New York: Chelsea House Publishers.

Langsdorf, W. (1936). *LZ-129 Hindenburg.* Bechold: Frankfurt.

Lehman, E.A. (1937). *Zeppelin.* London: Longmans Green.

Mather, M. (1937). I was on the Hindenburg. *Harper's* November.

Mooney, M.M. (1972). *The Hindenburg.* New York: Dodd, Mead & Co.

Neilsen, T. (1955). *The Zeppelin story.* London: Allen Wingate.

New York Times. (1937), Tuesday, 20th July.

O'Brien, P. (2000). *The Hindenburg.* New York: Henry Holt.

Payne, L. (1991). *Lighter than air: An illustrated history of the airship.* New York: Orion.

Robinson, D.H. (1964). *The LZ-129, Hindenburg.* Dallas, TX: Morgan Aviation.

Robinson, D.H. (1973). *Giants in the sky: A history of the rigid airship.* Seattle, WA: University of Washington Press.

Rosendahl, C. (1931). *Up ship.* New York: Dodd, Mead & Co.

Rosendahl, C.E. (1938). *What about the airship.* New York: Charles Scribner.

Sullivan, G. (1988). *Famous blimps and airships.* New York: Dodd, Mead and Co.

Tanaka, S. (1993). *The disaster of the Hindenburg.* Toronto: Madison Press.

Toland, J. (1957). *Ships in the sky.* New York: Henry Holt and Company.

Toland, J. (1972). *The great dirigibles: Their triumphs and disasters*. New York: Dover.

Vaeth, J.G. (1958). *Graf Zeppelin: The adventures of an aerial globe trotter*. New York: Harper.

CHAPTER 4: THE LARGEST MOVING OBJECT EVER BUILT

Arco Publishing (1970). *The White Star triple screw Atlantic liners Olympic and Titanic.* New York: Arco.

Ballard, R.D. (1987). *The Discovery of the Titanic.* New York, NY: Warner Books.

Ballard, R.D. (1988). *Exploring the Titanic: How the greatest ship ever lost-was found.* Toronto, Canada: Madison Press.

Bartlett, W.B. (2011). *Titanic: Nine hours to hell, the survivors' story.* Amberely Publishing: Stroud, England.

Beavis, D. (2002). *Who sailed on the Titanic? The definitive passenger list.* Hersham: Ian Atlan Ltd.

Beesley, L. (1912). *The loss of the SS Titanic,* Boston: Houghton-Mifflin.

Beesley, L., Gracie, A., Lightoller, C., & Bride, H., (1960). *The story of the Titanic as told by its survivors.* Wincour, J. (Ed.). New York: Dover.

Biel, S. (1996) *Down with the old canoe: A cultural history of the Titanic disaster.* New York: W.W. Norton.

Biles, J.H. (1912). The loss of the Titanic. *The Engineer, 113,* 409-410.

Bonsall, T.E. (1987).*Titanic.* New York: Gallery Books.

Booth, J. (1995). Two vital wireless messages that should have prevented the Titanic disaster. *The Titanic Commutator, 18,* 63.

Booth, J., & Coughlan, S. (1993). *Titanic: Signals of disaster.* Westburg, England: White Star Publications.

Braynard, F.O. (1988). *Story of the Titanic.* New York: Dover.

Bryceson, D. (1997). *The Titanic disaster: As reported in the British national press, April-July, 1912.* New York: W.W. Norton.

Bullock, S.F. (1913). *A Titanic hero: Thomas Andrews, shipbuilder.* Baltimore: Norman, Remington.

Cairis, N.T. (1992). *Era of the passenger liner*. Boston: Pegasus.

Carmichael, C. (1972). Was the Titanic 'Unsafe at any speed?' *Steamboat Bill*, Spring, 5-9.

Carrothers, J.C. (1962). The Titanic disaster. *US Naval Institute Proceedings, 88*, 57-69.

Cherry-Garrard, A. (1922). *The worst journey in the world*. New York: Carroll & Graf.

Committee on Commerce (1912). *The causes leading to the wreck of the White Star Liner Titanic: U.S. Senate Inquiry*. New York: U.S. Senate.

Conrad, J. (1912). Some reflexions, seamanlike and otherwise, on the loss of the Titanic. *The English Review II*, 304-315.

Conrad, J. (1912). Some aspects of the Admiralty inquiry. *The English Review II*, 581-595.

Croall, J. (1978). *Disaster at sea*. New York: Stein and Day.

Davie, M. (1986). *The Titanic: The full story of a tragedy*. London: Grafton Books.

Dodge, W. (1912). *The loss of the Titanic*. Riverside, CT: 7C's Press.

Donnelly, J. (1987). *The Titanic lost... and found*. New York: Random House.

Eaton, J.P., & Haas, C.A. (1987). *Titanic, destination disaster: The legends and the reality*. London: W.W. Norton.

Eaton, J.P., & Haas, C.A. (1994). *Titanic: Triumph and tragedy*. New York: W.W. Norton.

Everett, M. (1912). *Wreck and sinking of the Titanic*. New York: Russell.

Frey, S.B., Savage, A.D., & Torgler, B. (2011). Who perished on the Titanic? The importance of social norms. *Rationality and Society, 23*, 35-49.

Gardiner, R., & Van der vat, D. (1998) *The Titanic conspiracy*. Secaucus, New Jersey: Carol Publishing Group.

Gardner, M. (1986). (Ed.). *The wreck of the Titanic foretold.* (incl. Robertson 1898): Buffalo, NY: Prometheus.

Gelles, J.B. (1998). *Titanic: Women and children first.* New York: W. W. Norton & Co.

Gill, A. (2010). *Titanic: Building the world's most famous ship.* Guilford, CT: Lyon's Press.

Gracie, A. (1913). *The truth About the Titanic.* New York: M. Kennerley.

Haas, C. (1986). *Titanic: Triumph and tragedy.* Sparkford, England: W.W. Norton.

Hines, R.D. (2012). *Voyagers of the Titanic.* Harper Collins: New York.

Howells, R. (1999). *The myth of the Titanic.* New York, St. Martin's Press.

Lightoller, C.H. (1935). *Titanic and other ships.* London: Ivor, Nicholson, and Watson.

Lord, W. (1955). *A night to remember.* New York: Holt, Reinhart, and Winston.

Lord, W. (1987). *The night lives on.* New York, NY: Jove Books.

MacInnis, J. (1998). *Fitzgerald's storm: The wreck of the Edmund Fitzgerald.* San Diego, CA: Thunder Bay Press.

Marcus, G. (1969). *The maiden voyage.* New York: Viking Press.

Maxtone-Graham, J. (1997). *Titanic survivor: The newly discovered memoirs of Violet Jessup who survived both the Titanic and Britannic disasters.* Dobbs Farry, New York: Sheridan House.

Marshall, L. (1912). *Sinking of the Titanic and great sea disasters.* Philadelphia: J.C. Winston.

McCarty, J.H., & Foecke, T. (2008). *What really sank the Titanic.* New York: Citadel Press.

Merideth, L.W. (1999). *1912 facts about Titanic.* Sunnyvale, CA: Historical Indexes Publishing Co.

Mowbray, J.H. (1912). *Sinking of the Titanic.* Harrisburg, PA: The Minter Co.

National Geographic (1995). *The superliners: Twilight of an era.* WQED Pittsburgh (Video).

Neil, H. (1912). *Wreck and sinking of the Titanic.* Chicago: Homewood Press.

Padfield, P. (1965). *The Titanic and the Californian.* New York: John Day Co.

Pellegrino, C. (1988). *Her name, Titanic: The untold story of the sinking and finding of the unsinkable ship.* New York: Avon Books.

Pellegrino, C. (2000). *Ghosts of the Titanic.* New York: Morrow.

Poulton, E.C., Hitchings, N.B., & Brooke, R.B. (1965). Effect of cold and rain upon the vigilance of lookouts. *Ergonomics, 8,* 163-168.

Report on the Loss of the S.S. Titanic. (1912). St. Martin's Press: New York (1990).

Robertson, M. (1898). *The wreck of the Titan.* New York: Mansfield.

Ruffman, A. (2000). *Titanic remembered: The unsinkable ship and Halifax.* Toronto, Canada: Halifax Formac.

Russell, T.H. (1912). *Sinking of the Titanic.* Chicago: Homewood Press.

Scerbo, M.W. (1999). The RMS Titanic: A siren of technology. In: M.W. Scerbo and M. Mouloua (Eds.). *Automation technology and human performance.* (pp. 21-27), Erlbaum: Mahwah, NJ.

Spufford, F. (1997). *I may be some time.* New York: St. Martin's Press.

Stenson, P. (1954). *The odyssey of C.H. Lightoller.* New York: W.W. Norton.

Thayer, J.B. (1940). *The sinking of the S.S. Titanic.* Riverside, CT: 7C's Press.

U.S. Congress Senate (1912). Hearings of a Subcommittee of the Senate Commerce Committee pursuant to Senate Resolution 238, to Investigate the Causes Leading to the Wreck of the White Star Liner, Titanic, 62nd Congress, 2nd Session, 1912, Senate Document 726 (#6167), Washington, DC.

U.S. Congress Senate (1912). Report of the Senate Committee on Commerce pursuant to Senate Resolution 238, Directing the Committee to Investigate the Causes of the Sinking of the Titanic, with speeches by William Alden Smith and Isador. Rayner, 62nd Congress, 2nd Session, 28th May, 1912, Senate Report 806 (#6172), Washington, DC.

U.S. Congress Senate (1912). *Loss of the steamship Titanic: Report of a Formal Investigation as conducted by the British Government.* Presented by Mr. Alden Smith. 62nd Congress, 2nd Session, 20th August, 1912, Senate Document 933 (#6179), Washington, DC.

Walker, B.J. (1912). *An unsinkable Titanic.* New York.: Dodd, Mead, & Co.

Wade, W.C. (1979). *The Titanic: End of a dream.* New York: Penguin.

Wels, S. (1997). *Titanic: Legacy of the world's greatest ocean liner.* New York: Time-Life.

Winocur, J. (1960) (Ed.). *The story of the Titanic: As told by its survivors.* New York: Dover.

Wreck Commissioner's Court (1912). *Formal investigation into the loss of the S.S. Titanic: Evidence, appendices, and index.* Reprinted 1998. London: Public Records Office.

CHAPTER 5: REACHING FOR GOD

Adams, H. (1928). *Mont-Saint-Michel and Chartres*. Boston: Houghton-Mifflin.

Anonymous (1989). *Notre-Dame de Chartres*. IGP.

Arms, D.N., & Arms, J.T. (1929). *Churches of France*. New York: MacMillan.

Aubert, M. (1965). A propos de l'eglise abbatiale de Saint-Lucien de Beauvais. In: *Gedenkschrift Ernst Gall* (M. Kuhn and L. Grodecki [Eds]). (pp. 51-58). Berlin and Munich. Deutscher Kunstverlag.

Aubert, M., & Goubert, S. (1959). *Gothic Cathedrals of France and their treasures*. London: N. Kaye.

Bernat, R.A. (1992). *History of the building of Beauvais Cathedral and the collapse of the Choir in 1284*. Senior Thesis. Colorado College, Department of Art.

Blois, V. (undated). *Chartres: The Cathedral, The Town, The old quarters, The main monuments of Eure-et-loir*. Chartres: Valoire Publishing.

Bony, J. (1983). *French gothic architecture of the 12th and 13th centuries*. Berkeley, CA: University of California Press.

Branner, R. (1962). Le maitre de la cathedrale de Beauvais. *Art de France, II*, 77-92.

Branner, R. (1962). The labyrinth of Reims Cathedral. *Journal of the Society of Architectural Historians, XXI*, 18-25.

Cormack, P. (1984). *English cathedrals*. New York: Harmony Books.

Clifton-Taylor, A. (1967). *The cathedrals of England*. London: Thames & Hudson.

Duby, G. (1981). *The age of cathedrals: Art and society 980-1420*. Chicago: University of Chicago Press.

Edwards, D.L. (1989). *The Cathedrals of Britain*. Andover, England: Pitkin Pictorials.

Fitchen, J. (1961). *The construction of Gothic Cathedrals: A study of medieval Vault erection*. Oxford. Clarendon Press.

Forstel, J., & Magnien, A. (2005). *Beauvais Cathedral*. Picardy Dept. of Culture: Amiens.

Fosca, F. (1959). Introduction. In: R. Jacques, *French Cathedrals*. Munich: Wilhelm Andermann Verlag.

Gimpel, J. (1959). *Les bâtisseurs de cathédrales*. Paris: Éditions du Seuil.

Heyman, J. (1967-1968). Beauvais Cathedral. *Transactions of the Newcomen Society, XL*, 15-32.

Jacques, R. (1959). *French cathedrals*. Wilhelm Andermann: Munich.

Macaulay, D. (1973). *Cathedral: The story of its construction*. Boston: Houghton-Mifflin.

Mark, R. (1972). The structural analysis of Gothic Cathedrals. *Scientific American, 227 (5)*, 90-99.

Miller, M. (1981). *Chartres: The cathedral and the old town*. Andover, Hants, England: Pitkin Pictorials.

Miller, M. (undated). *Chartres: Guide of the Cathedral*. Houvet-La Crypte: Notre Dame.

Murray ,S. (1976). *The collapse of 1284 at Beauvais Cathedral*. Acta, III, 17-44.

Murray ,S. (1989). *Beauvais Cathedral: Architecture of transcendence*. Princeton, NJ: Princeton University Press.

Planel, P. (1991). *Old Sarum: A handbook for teachers*. English Heritage: White Crescent Press.

Robert, J. (1989). *Notre-Dame de Chartres*. Chartres: Cathedral Publications.

Swaan, W. (1984). *The gothic cathdral*. Park Lane: New York.

Wolfe, M.I., & Mark, R. (1976). The collapse of the vaults of Beauvais Cathedral in 1284. *Speculum, LI*, 462-476.

❖ ❖ ❖

CHAPTER 6: SURVIVING SISTERS

Adams, H. (1986). *Mont Saint Michel and Chartres*. Penguin: New York.

Adams, J.L. (1991). *Flying buttresses, entropy and O-rings*. Cambridge, MA: Harvard University Press.

Ball, P. (2008). *Universe of stone: A biography of Chartres Cathedral*. New York: Harper.

Bonsall, T.E., & White Star Line. (1987). *Titanic: The story of the great White Star Line trio; the Olympic, the Titanic and the Britannic*. New York: Gallery Books.

Dick, H.G., & Robinson, D.H. (1984). *The golden age of the great passenger airships: Graf Zeppelin & Hindenburg*. Washington, DC: Smithsonian Institution Press.

Duby, G. (1981). *The age of the cathedrals: Art and society 980-1420*. Chicago, IL: University of Chicago Press.

Eaton, J.P., & Haas, C.A. (1990). *Falling star: Misadventures of White Star Line ships*. New York: W.W. Norton.

Eckner, H. (1949). *Graf Zeppelin*. J.G. Cott'sche Buchandlung. Nachfolger: Stuttgart.

Edwards, D.L. (1989). *The cathedrals of Britain*. Andover, England: Pitkin.

Erlande-Brandenburg, A. (1993). *The cathedral builders of the Middle Ages*. London: Thames & Hudson.

Faulkner, M. (2012). *The Kriegsmarine and the Aircraft Carrier: The design and operational purpose of the Graf Zeppelin, 1933-1940. War in History, 19(4)*. 492-516.

Fenwick, V., & Gale, A. (1998). *Historic shipwrecks: Discovered, protected, and investigated*. Tempus: Stroud.

Fleming, J.A. (1917). *The last voyage of His Majesty's hospital ship "Britannic"*. London: Marshall.

Gardiner, R. (2001). *The history of the White Star Line*. Hersham: Ian Allan.

Jantzen, H. (1962). *High Gothic: The classic cathedrals of Chartres, Reims, Amiens.* New York: Pantheon.

Lehmann, E., & Mingos, H. (1927). *The Zeppelins.* New York: J.H. Sears & Co.

Markale, J. (1988). *Cathedral of the black Madonna: The druids and the mystery of Chartres.* Rochester, VT: Inner Traditions.

Marshall, L., & University of Virginia. (1996). S*inking of the Titanic and great sea disasters.* Charlottesville, VA: University of Virginia Library.

Miller, M. (1980). *Chartres Cathedral.* Andover, England: Pitkin Pictorials.

Miller, W.H. (1988). *Famous ocean liners.* New York: Dover.

Norway, N.S. (1954). *Slide rule Autobiogrophy of an engineer.* London: Heinemann.

Syon, G.. (2002). *Zeppelin!: Germany and the airship, 1900-1939.* Baltimore: Johns Hopkins University Press.

Toman, R. (2004). (Ed.). *Gothic.* Tandem: Goodfellow and Egan Publishers, Cambridge.

Vaeth, J.G. (2005). *They sailed the skies: U.S. Navy balloons and the airship program.* Annapolis, MD: Naval Institute Press.

Wintringham, T.H. (1930). The crime of the R-101. *Labour Monthly,* December.

CHAPTER 7: THE RIDDLE OF THE LABYRINTH

Branner, R. (1962). Labyrinth of Reims cathedral. *Journal of the Society of Architectural Historians,* 2118-25.

Compton, V. (2007). *Understanding the Labyrinth as transformative site, symbol, and technology: An arts-informed inquiry.* Toronto, Canada: ProQuest.

Connolly, D. K. (2005). At the center of the world: The labyrinth pavement of Chartres Cathedral. *Studies in Medieval and Reformation Traditions*, 285. Brill Academic Publishers: Boston, MA.

Doob, P. (1990). *The idea of the labyrinth from classical antiquity through the Middle Ages*. Ithaca: Cornell University Press.

Ferré, R. (2001). *Church labyrinths: Questions & answers regarding the history, relevance, and use of labyrinths in churches*. St. Louis, MO: One Way Press.

Fisher, A. (2004). *Mazes and labyrinths*. Shire: Princes Risborough, England.

Fisher, A., & Gerster, G. (2000). *The art of the maze*. London: Phoenix Illustrated.

Fox, M. (2013). *The riddle of the labyrinth: The quest to crack an ancient code*. Harper-Collins: New York.

Gifford, D. (1990). *The further shore: A natural history of perception*. New York: Atlantic Monthly.

James, J. (1977). Mystery of the great labyrinth Chartres Cathedral. *Studies in Comparative Religion, 11*(2), 92-115.

Kern, H. (2000). *Through the labyrinth: Designs and meanings over 5000 years*. Hermann Kern. Munich; New York: Prestel.

Matthews, W. (1970). *Mazes and labyrinths; their history and development*. New York: Dover Publications.

McCullough, D.W. (2004). *The unending mystery: a journey through labyrinths and mazes*. New York: Pantheon Books.

Morrison, T. (2003). The labyrinthine path of pilgrimage. *Peregrinations. Journal of Medieval Art and Architecture, 1*, 1-7.

Mortimer, R. (1991). *A hidden treasure revealed: An introduction to the Cosmati Pavement*. Westminster Abbey Occasional Papers, Second Series No. 1.

Olsen, O. (2010). *Temple Antiquities: The templar papers II*. Ropley, England: O Books

Saward, J. (2003). *Labyrinths & mazes: A complete guide to magical paths of the world*. Asheville, NC: Lark Books

Westbury, V. (2003) *Labyrinths ancient paths of wisdoms*. Cambridge, MA: De Capo Press

Wilson, F., & Bancroft-Hunt, N. (1996). *Labyrinth and maze*. Oxford University Press

Wright, C. M. (2001). *The maze and the warrior: Symbols in architecture, theology, and music*. Cambridge, MA: Harvard University Press.

CHAPTER 8: SHIPS OF THE SOUL

Aristotle (c. 326 BC). *Poetics*. (Trans. M. Heath, 1996), London: Penguin.

Barber, P. (1988). *Applied cognitive psychology*. London: Methuen.

Burton, R., & Cavendish, R. (1991). *Wonders of the world*. Chicago: Rand-McNally.

Gerber, P. (1997). *Stone of destiny*. Edinburgh: Canongate Books.

Gibson, J.J. (1958). Visually controlled locomotion and visual orientation in animals. *British Journal of Psychology, 49*, 182-194.

Gibson, J.J. (1979). *The ecological approach to visual perception*. Boston: Houghton-Mifflin.

Hancock, P.A. (2007). On time and the origin of the theory of evolution. *Kronoscope, 6* (2), 192-203.

Macaulay, D. (1973). *Cathedral: The story of its construction*. Boston: Houghton-Mifflin.

Murphy, M. (1972). *Golf in the kingdom*. London: Penguin.

Schiff, W.(1965). Perception of impending collision: A study of visually directed avoidant behavior. *Psychological Monographs, 79*, Whole No. 604.

Schiff, W., Caviness, J.A., & Gibson, J.J. (1962). Persistent fear responses in Rhesus Monkeys to the optical stimulus of 'Looming.' *Science, 136*, 982-983.

Schiff, W., & Detweiler, M.L. (1979). Information used in judging impending collision. *Perception, 8*, 647-658.

Von Simson, O. (1956). *Gothic cathedral: Origins of Gothic architecture and the medieval concept of order*. New York: Pantheon Books.

❖ ❖ ❖

CHAPTER 9: THREADS THROUGH TIME

Anderson, J. (2014) *Airship on a shoestring the story of R100*. Gamlingay, England: Authors OnLine Ltd.

Bennett, G. (1961). *By human error: Disasters of a century*. London: Seeley Service & Co.

Chamberlain, G. (1984). *Airships: Cardington*. Dalton Ltd; Lavenham, UK.

Christopher, J. (2010). *Transatlantic airships: An Illustrated History*. Marlborough, England: Crowood Press.

Countryman, B. (1982). *R100 in Canada*. Ontario, Canada: Boston Mills Press.

Deighton, L., & Schwartzman, A. (1979). *Airshipwreck*. New York: Holt, Rinehart, & Winston.

Dorner, D. (1996). *The logic of failure: Recognizing and avoiding error in complex situations*. New York: Metropolitan Books.

Eberhart, M.E. (2003). *Why things break: Understanding the world by the way it comes apart*. New York: Harmony Books.

Fuller, J.G. (1979). *The airman who would not die*. New York: Putnam & Sons.

Gilbert, J. (1976). *The world's worst aircraft*. New York: St Martin's Press.

Griehl, M., & Dressel, J. (1990). *Zeppelin! The German airship story*. London: Arms and Armour.

Hallion, R. P. (2010). Ethereal dreams of imperial airships: R 101's ill-fated final flight brought Britain's long-range dirigible program to a tragic conclusion. *Aviation History*, (6). 52.

Hartcup, G. (1974). *The achievement of the airships: A history of the development of rigid, semi-rigid and non-rigid airships*. Newton Abbott, England: David & Charles.

Harvey, J. (1950). *English cathedrals*. London: Batsford.

Icher, F. *(1998). Building the great cathedrals.* (Translated from the French by Anthony Zielonka). New York: Harry N. Abrams.

Job, M. (1996) *Air disaster.* Fyshwick, Australia: Aerospace Publications.

Leasor, J. (1957). *The millionth chance: The story of the R-101.* New York: Reynals & Co.

Mallan, L. (1962). *Great air disasers.* New York: Faucett.

Masefield, P.G. (1982). *To ride the storm.* London: Kimber.

Morpurgo, J. E. (1972). *Barnes wallis: A biography.* Berkeley, CA: St. Martin's Press.

Report of the R-101 Inquiry (1931). *R-101 The airship disaster, 1930.* London: HMSO.

Swinfield, J. (2011). Egos and aeronautics: A tale of two airships. *History Today, 61* (6), 26-29.

Wintringham, T. H. (1930). *Crime of R101. Labour Monthly,* 12733-738.

CHAPTER 10: TRANSPORTS OF DELIGHT

Atkinson, R.J.C. (1960). *Stonehenge.* London: Pelican.

Balfour, M. (1980). *Stonehenge and its mysteries.* New York: Scribner.

Burgess, C. (1980). *The age of Stonehenge.* London: J.M. Dent & Sons.

Burl, A. (1991). *Prehistoric henges.* Shire Publications: Princes Risborough.

Burl, A. (1996). *Stonehenge.* Oxford: Oxford University Press

Burl, A. (2000). *The stone circles of Britain, Ireland, and Brittany.* New Haven: Yale University Press.

Burl, A. (2001). Stonehenge: How did the stones get there? *History Today, 51* (3), 19-25.

Burl, A. (2007). *A brief history of Stonehenge: One of the most famous ancient monuments in Britain.* New York, NY: Running Press.

Burl, A. (1999). *Great stone circles: Fables, fictions, facts.* New Haven, CT: Yale University Press.

Chippindale, C. (1983). *Stonehenge complete.* Ithaca, NY: Cornell University Press.

Eberhart, M.E. (2003). *Why things break: Understanding the world by the way it comes apart.* New York: Harmony Books.

Hawkins, G. S. (1973). *Beyond Stonehenge.* New York: Harper & Row.

Hawkins, G.S., & White, J.B. (1965). *Stonehenge decoded.* Garden City, NY: Doubleday.

Hill, R. (2008). *Stonehenge.* Cambridge, MA: Harvard University Press.

Hoyle, F. (1977). *On stonehenge.* London: Heinemann Educational.

Jahn, R.G., & Dunne, B.J. (1987). *Margins of reality: The role of consciousness in the physical world.* San Diego, CA: Harcourt, Brace Jovanovich.

Johnson, A. (2008). *Solving Stonehenge: The new key to an ancient enigma.* London: Thames & Hudson.

Langdon, R.J. (2014). *The Stonehenge enigma (Prehistoric Britain).* East Sussex, England. ABC Publishing Group.

Leasor, J. (1957). *The millionth chance: The story of the R-101.* New York: Reynals & Co.

Masefield, P.G. (1982). *To ride the storm: The story of the airship R-101.* London: Kimber & Co.

Newham, C.A. (1972). *The astronomical significance of Stonehenge.* John Blackburn: Leeds

Pearson, M.P. (2013). *Stonehenge—A new understanding: Solving the mysteries of the greatest stone age monument.* New York, NY: Workman Publishing.

Renfrew, C., & Cunliffe, B.W. (1997). *Science and Stonehenge*. Oxford: Oxford University Press.

Richards, J.C. (2005). *Stonehenge*. United Kingdom: English Heritage.

Richards, J.C. (2007). *Stonehenge: The story so far*. United Kingdom: English Heritage.

Stevens, F. (1933). *Stonehenge today & yesterday*. London: HMSO.

Thomas, H.H. (1923). The source of the stones of Stonehenge. *The Antiquaries Journal*, 3(03), 239-260.

CHAPTER 11: AUTOBIOMIESIS AND ULTIMATE PURPOSE

American Psychiatric Association. (2013). *Diagnostic and statistical manual of mental disorders* (5th ed.). Arlington, VA: American Psychiatric Publishing.

Aristotle, Cooke, H. P., Longinus, & Demetrius, (1926). *Aristotle.* Cambridge, Mass: Harvard University Press.

Barrett, P.M. (2012). *Glock: The rise of America's gun.* New York: Broadway.

Campbell, J. (1988). *The power of myth.* New York: Anchor.

Clarke, A.C. (1962). Hazards of prophecy: The failure of imagination. In: *Profiles of the future: An inquiry into the limits of the possible.* New York: Harper & Row.

Clark, K. (1969). *Civilization.* New York: Harper & Row.

Dostoyevsky, F. (1880). *The brothers Karamazov* (1952 Edition), Raleigh, NC: Spartan Press.

Eccles, J., & Robinson, D.N. (1985). *The wonder of being human.* New Science Library: Boston.

Gini, C. (1912). Variabilita and multibilita. *Stdui economico-Giuridici dell'Univ. di Calgliari. 3* (2) 1-158.

Gottleib, I.M. (1987). *Understanding oscillators.* Blue Ridge, PA: Tab Books.

Gregory, R.L. (1981). *Mind in science: A history of explanation in psychology and physics.* New York: Cambridge University Press.

Hancock, P.A. (2009). *Mind, machine and morality.* Chichester, England: Ashgate

Hancock, P.A. (2012). *The making of myth: Edward Leedskalnin and the Coral Castle. Skeptic, 18*(1), 44-50

Hancock, P.A. (2013). Driven to distraction and back again. In: M.A. Regan T, T. Victor, and J. Lee, (Eds.). *Driver distraction and inattention: Advances in research and countermeasures* (pp. 9-25), Ashgate, Chichester, England.

Hancock, P.A. (2013). In search of vigilance: The problem of iatrogenically created psychological phenomenon. *American Psychologist*, 68 (2), 97-109

Hancock, P.A., Billings, D.R. & Oleson, K.E. (2011). Can you trust your robot? *Ergonomic in Design, 19* (3), 24-29

Hancock, P.A., Billings, D.R. Olsen, J.Y.C., de Visser, E.J. & Parasuraman, R, (2011). A meta-analysis of factors impacting trust in human-robot interaction. *Human Factors, 52* (5), 517-527

Hancock, P.A., Mouloua, M. & Senders, J.W. (2008). *On philosophical foundations of driving distraction and the distracted driver*. Boca Raton, FL: CRC Press.

Illich, I. (1973). *Tools for conviviality*. New York: Harper & Row.

Kauffman, S. (1993). *The origins of order*. Oxford: Oxford University Press.

Lee, J.D., & See, K.A. (2004). Trust in automation: Designing for appropriate relation. *Human Factors, 46* (1), 50-80.

Maslow, A.H., & Herzeberg, A. (1954). Heirarchy of Needs. In: A.H. Maslow. (Ed.) *Motivation and personality*. Harper, New York.

Parasuraman, R. & Riley, V. (1997). Humans and automation: Use, misuse, disuse, abuse. *Human Factors, 39* (2), 230-253

Regan, M.A., Lee, J.D., and Young, K.L. (Eds.) *Driver distraction: Theory, effects and mitigation*. Boca Raton, FL: CRC Press.

Rothenberg, D. (1993). *Hand's in: Technology and the limits of nature*. Berkeley: University of California Press.

Sarter, N. & Woods, D.D. (1995). How in the world did we ever get Into that mode? Mode error and awareness in supervisory control. *Human Factors, 37* (1), 5-19.

Searle, J. (1984). *Minds, brains, and science*. Harvard University Press: Cambridge, MA.

Smith, K., & Hancock, P.A. (1995). Situation awareness is adaptive, externally-directed consciousness. *Human Factors, 37* (1), 137-148.

Past Publications

Hancock, P.A., & Szalma, J.L. (Eds.). (2008).*Performance Under Stress.* Ashgate Publishing. Aldershot England.

Hancock, P.A. (2009). *Mind, Machine, and Morality.* Ashgate Publishing, Aldershot, England.

Hancock, P.A. (2009). *Richard III and the Murder in the Tower.* The History Press, England. (Papeback Edition, February 2011)..

Hancock, P.A. (2011). *Cognitive Differences in the Ways Men and Women Perceive the Dimension and Duration of Time: Contrasting Gaia and Chronos.* Edwin Mellen Press: Lewiston, New York.

Matthews, G., Desmond, P.A., Neubauer, C., & Hancock, P.A. (Eds.), (2012). *The Handbook of Operator Fatigue.* Farnham, Surrey, UK: Ashgate Publishing.

Hoffman, R.R., Hancock, P.A., Parasuraman, R., Szalma, J.L., & Scerbo, M. (2015). (Eds.). *Handbook of Applied Perception Research.* Cambridge, Cambridge University Press,

Hancock, P.A. (2015). *Hoax Springs Eternal: The Psychology of Cognitive Deception.* Cambridge: Cambridge University Press.

Printed in the United States
By Bookmasters